OUT OF PRINT
D0262655

THE SPICE OF LIFE

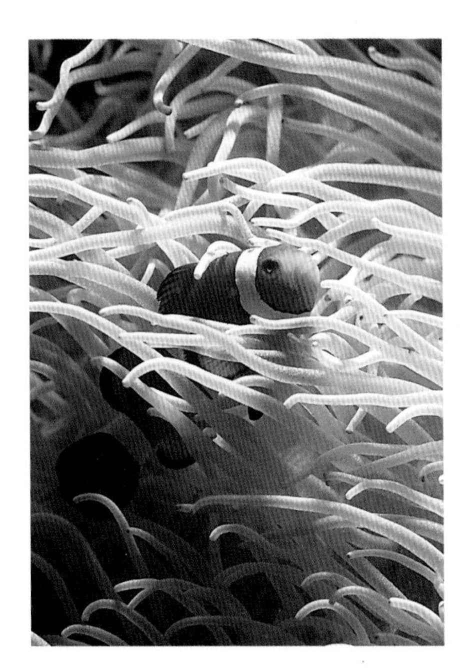

THE SPICE OF LIFE

BIODIVERSITY AND THE EXTINCTION CRISIS

CHRIS HOWES

BLANDFORD

To Smokey Joe, for sharing
more than a couple of lives

A BLANDFORD BOOK

First published in the UK 1997 by Blandford
A Cassell Imprint
Cassell Plc, Wellington House,
125 Strand, London WC2R 0BB

Distributed in the United States by Sterling Publishing Co., Inc.
387 Park Avenue South, New York, NY 10016-8810

British Library Cataloguing-in-Publication Data
A catalogue entry for this title is available from the British Library

ISBN 0-7137-2390-4

Designed by Richard Carr
Printed and bound by Kyodo Printing Co., Singapore.

CONTENTS

#

ACKNOWLEDGEMENTS

As with all good ideas, there is a beginning. In the case of *The Spice of Life*, that spark came from a good friend, Phil Chapman. Phil is a superb naturalist (a walk with him along a forest path or disused railway line, where the passage of hours is allowed to drift to infinity, is a stimulating experience), but never found time, in a pressing schedule of making wildlife films, to tackle the immensity of life's story. I was both honoured and touched when he asked me if I would take up the challenge.

As well as Phil I must also thank Cassell's natural history editor, Stuart Booth; his always prompt support and faith in the project have been of immense aid. Equally, there are many other friends who have helped along the way. In particular, I would like to record my grateful thanks to Anne Bell, Julie Davies, Rhian Hicks, Alan Jones, Andy Kendall, Jane Mitchell, Tom Sharpe and Jennifer Stewart. Added to this list must be draft reader and text pilot Alex Laxton and Mark Lumley for drawing the diagrams.

Acknowledging the author's partner seems commonplace in modern books. This is for a reason: no author engrossed in a major project could manage to stay sane without support and help. In my case I must thank Judith Calford, not only for helping with research and commenting on drafts, but also for simply being there.

Chris Howes
Cardiff
January 1996

PICTURE CREDITS

Each chapter is illustrated with a line drawing which, for the most part, originally appeared in Pouchet's *The Universe*, first published in France in 1865. The drawings depict, in chapter order: a flying ant (introduction); 'Octopus horridus' from *Cassell's Natural History* of 1881; a 'climbing perch'; a section through a nautilus shell; the common garden snail (*Helix aspersa*); the Pine Silk-worm Moth; a *Voluta* shell; the fossilized remains of a dragonfly; a white lily; a mole cricket; and, perhaps somewhat poignantly, a lemming in the epilogue.

All photographs are by the author, with the exception of those on pages 58, 107, 140, 168 and 169, which are by courtesy of Philip Chapman.

MEASUREMENTS

Measurements in this book are based on the metric system, with the exception of data in historical quotations. Reference to a billion represents a unit of one thousand million. To convert from the metric system to imperial measurements, the following conversion factors provide good approximations:

DIMENSIONS
centimetres to inches: × 0.39
metres to feet: × 3.3
kilometres to miles: × 0.62

TEMPERATURE
Celsius to Fahrenheit: × 1.8, then add 32

VOLUME
litres to pints: × 1.75

WEIGHT
grammes to ounces: × 0.035
kilogrammes to pounds: × 2.2

THE GODS OF GAIA

HERE ARE THE GODS surrounded with their celestial chessboard. There is Thor, who, armed with a mighty hammer, broke up earth's surface to fashion mountains from the splinters. Here is Pan-Kou-Ché, an old man armed with hammer and chisel, the Chinese creator god. Ra, Egyptian ruler of the earth, looks disinterested.

One chess piece is moved through time and space by Yggdrasil, the great world-tree; the bidding starts.

How much for this planet? An ideal, comfortably warm blue-green world, fully furnished with life. We begin at thirty million species. Now twenty-five million . . . twenty million? Will no one make an offer? We are now accepting bids at fifteen million species. Gods and goddesses, ladies and gentlemen, we are approaching our reserve. Is there no value in the remaining ten million?

Yes, a question? Well, agreed, there have been distasteful alterations by the dominant species, but it still represents a valuable commodity in the firmament. Come along now, we're asking for a realistic assessment. The loss of life is taken into account. The investment is still good, but the value is decreasing; who will make a bid for the remaining five million species, declining? It seems that no one is interested. Can this be true? Shall the reign of the current dominant life-form continue; is it worth saving? Do not upset our friend Gaea further; she already weeps. Place a bid, please.

No? We have wasted enough time and this world becomes devalued. We must move to the next lot. I am forced to raise my hammer over life on earth. Your last chance to bid for these dwindling species . . .

Going, going . . .

PREFACE

TAKE A RICH mixture of life, a gentle squeeze of evolution, add a pinch of history and flavour liberally with science. Allow to ferment, and serve at world temperature. The result is *The Spice of Life*, a study of how life began, what affects its species, and where it is heading.

When I embarked upon this project into biodiversity and the extinction crisis – what caused life to arise, how is it formed, how is it threatened – I little recognized how far reaching the research would be. In particular, it encompassed such a broad band of subject matter that it was difficult to know where to start and end: is *this* research worthy of mention, does *that* theme merit inclusion? If truth is told, the topic is vast. Theories conflict and new perceptions spring forth with the regularity of a sunrise; where one idea begins to find acceptance, another is created to tear down the pedestal. Science works to check and recheck, to imagine and oppose, to take leaps of intellect across a sea of contrasts until a hypothesis is eventually accepted as fact, before being challenged and the process beginning again.

The problem with biodiversity is that it is . . . well, diverse. Complete volumes have been written on the theories which appear as the minutia of *Spice*; what is the truth, and what is not? Which conflicting opinion is the more valid of two, or four, or a hundred? Life, with respect to time, has an unimaginable history; we can only infer and conjecture as to how it began. Each question eventually procures an answer, which creates new questions to spawn more theories and . . .

Science is not omnipotent and, of its disciplines, ecology deals not in tangible neutrons and chemicals which *always* join just so. Little by little, we learn more of our world through the study of life, which follows its own rules. Why did the dinosaurs die? We wonder what prompted the demise of a species when we do not know details of its metabolism, its behaviour, or even the colour of its skin. As the next millennium approaches like a speeding train, we threaten our modern biodiversity. Do clues lie with dinosaur fossils which might help man's survival? How do populations interreact? We follow a road through time with hills and bends, forks and crossroads, and must decide the route to take.

The Spice of Life offered me an opportunity to think of where life began, of what forces affected its evolution and where science is taking us. Of how man initiates new forms of life and patents his own genes. By what means we influence our world, of our conservation and pollution. To consider how little we truly know, and therefore how much we rely on unproved and conflicting theories. Of where we should head next, to sway the course of our future.

As I researched this book, I discovered how quickly the world is undergoing change and how urgent the need to initiate controls. *The Spice of Life* is a story of enchantment, of charm, beguilement and rapture with the forces of the natural world. *The* natural world: it's the only one we have. It's time we recognized that one, inalienable fact.

INTRODUCTION

IMAGINE

When we no longer look at an organic being as a savage looks at a ship, as something wholly beyond his comprehension; when we regard every production of nature as one which has had a long history; when we contemplate every complex structure and instinct as the summing up of many contrivances, each useful to the possessor, . . . how far more interesting – I speak from experience – does the study of natural history become!

Charles Darwin, *Origin of Species*, 1859

IMAGINE A DARKENED, brown room. Heavy drapes are held each side of a north-facing window, sunlight wanly peering in from a cobbled street. There is a slightly musty smell, as if the room requires airing more often. The slow tock, tick, tock of a grandfather clock paces the minutes. A large, oak table, perfectly polished, dominates the carpet, while walls bear a writing bureau, books bound in calf and cloth, and a series of wooden cabinets. A preserved owl, a few small bones neatly numbered, snail shells, racks of insects pinned to cork in serried rows. The tools of the trade lie to one side: a butterfly net, pins, killing jar, a microscope near the window. This is a room of the nineteenth century, dedicated to natural history: a collector's room.

How precise is this stereotyped image, fostered by Hollywood? There certainly were Victorian collectors who ranked elements of the natural world one above the other and performed endless, minute comparisons of beetles and moths, grasses and flowers. Why? What drove them to such dedication? What did these often amateur naturalists hope to achieve?

The answer is not as altruistic as might be first assumed; there were sometimes other motives in force than simply a search for knowledge and science, such as entertainment. Neither is this Victorian world the start of the search for and recording of biological diversity: biodiversity. That story starts far earlier.

There is an immense catalogue of data in existence which covers all the known species of earth, with more being continually added. This is the arena of biological specialists: ask any museum or specialist for a list of all known insects and you will receive a blank look. Try for the latest species' names, some of which have changed once, twice, and more times in recent years, and the same response is likely. Yes, more species are continually added as collections are made, differences are noted, and the catalogue changes – but there are so many niches for study that it is impossible to follow every minute addition. It is no longer possible for any one person to keep up with all the literature (there are now 500,000 scientific journal titles listed by UNESCO), any more than a doctor can read every medical journal and know details of every new drug. New species are being discovered daily; yet more life to name and file. While the rare and unexpected discovery of a large mammal new to science causes waves of surprise around the world, more microbes, worms and beetles are ten a penny.

The cycle of life is never ending: leaves take in sunlight and grow on the forest floor, nourishing the plant until it dies, to decay under the forces of fungi and decomposers, leaving its nutrients available for the next generation.

A true and complete catalogue of life? There is no such thing. There is only a feeling that we have been through the forests and oceans of earth and catalogued what we have found, however poorly. So how do we know how many species there are on this planet? We don't – we can only make estimates, just as we are restricted to estimating the numbers of living, individual organisms. Some estimates are more accurate than others, of course, and we find interesting facts: the largest single group of organisms is that of the insects, with over three-quarters of a million species described. As a group, this means that insects outnumber mammalian species at something approaching 200:1, and all plants by about 3:1. Through all time, insects have been the most successful of all life-forms.

It goes further. Choose a number for the extinction rate of animals and plants: how many species wink out of existence, to join the world of fossils, for all the good their genes brought them? Is it ten a year, ten a week, ten a day? How about the staggering estimate of four every hour, day and night, year in year out, a rate of extinction that far outstrips all others throughout the millennia. Fossils themselves represent a fixed record of life that was; why did the organisms die, what caused the loss of their species? How many animals and plants have existed in the distant past, to go the way of the dinosaurs? What of evolution and the production of all this life? How does it work?

It's a numbers game to form arguments like this, serving only to fuel the statistician's pen. Biodiversity is a wider concept, a word which represents the array of living things, known and unknown. There is a real world of nature out there, one which numbers try to quantify. Biodiversity struggles to exist with or without that expression; it is a blind lottery of chance which controls the progression. You are reading this volume so you already, presumably, have an interest in this living world, a curiosity of passing importance or driving force. Others – friends, neighbours, 'foreigners', businessmen, media presenters, politicians – may feel differently.

It comes down to this: why does life exist on our earth, yet other planets remain barren? What is life; how did it form, how does it die? How do we know how many living organisms and species there are? Of even more importance, how do we estimate the numbers yet to be found, identified and catalogued? How do we know where to direct our energies into the study of biodiversity and, once we begin to find answers, what do we do with them? We have knowledge of these life-forms, this biodiversity. We, the species supposedly at the pinnacle of an evolutionary procession, can count, we can measure. We question.

Does it stagger you that 99 per cent of all species which have ever existed are now extinct? That some hundred or so species now follow them into oblivion every day? Does it matter? So what?

Who cares?

Jewel anemones waiting to ensnare prey.

CHAPTER ONE

MAPPING A WORLD OF DISORDER

Nature is simply the system of laws established by the Creator for the existence of things and the multiplication of human beings. Nature is not herself a thing, for the thing would then comprehend the universe; Nature is not a being, for the Being would be the Almighty; but she must be regarded as an active power, immense, limitless, embracing every thing, animating every thing, but subordinated to the will of the Great Being.

Buffon, *Nature*, 1866

IN EVERY WALK of modern life you meet specialists: builders, dentists, farmers, scientists – or even tinkers, tailors, soldiers, spies. Look deeper, and more specialities emerge: farmers keep sheep or goats, and grow maize or beans. The same is true for science; is your specialist a physicist, a chemist or a biologist? Does the chemist study organic or inorganic chemistry, the biologist the life of a saltmarsh or the operation of an eye? Already, a system of grouping has occurred; there is nothing mystical here, just the ability to place similar disciplines together. What defines a biologist? A person who studies life. What defines a geneticist? A biologist specializing in the study of genes and their actions. A biochemist? A worker in the field of biological chemistry. And so on; the specialists have been classified.

To place things – objects or concepts – into groups is very basic; we teach our children to do so, piling up play bricks or matching like colours with like. Grouping helps us learn to deal with a complex world by making it manageable. Few groups are definitive, though; how many times have you hit upon the 'odd one out' or 'place into groups' problem?

Question. Which is the odd one out from: stool, chair, table, bench?

Question. Divide into two groups, each with something in common: knife, fork, plate, spoon, cup.

Most people will choose the table as the odd one out, as you can sit on all the others, and group the knife, fork and spoon together as they are all things you eat with (or because they are made of metal while the others in the group are not). What happens, then, when the questions are answered with 'stool' and 'spoon, plate, cup', because a stool has three legs, not four, and a spoon, plate and cup can all hold a liquid? In a rigid framework of an answer sheet, these answers may not be acceptable; all the answers assume prior knowledge, such as the use of the object or the material it is made with. In the same manner, classifying scientific specialists may not be straightforward or yield acceptable answers: what of the biologist studying the physics of movement, or the chemical reaction to light in the eye? Is this now pure biology, physics, chemistry, or some as yet unclassified hybrid?

The point is that grouping things together may not be straightforward and, if it gives difficulties even with these small examples (which themselves form groups), what problems must exist within our modern system of biological classification? The system is not perfect, but it does contain the disciplines of taxonomy (the word derives from the Greek *taxis*, meaning 'arrangement', and is involved with grouping and naming organisms: nomenclature) and systematics (revealing evolutionary relationships).

The story of classification goes back to the closing centuries before Christ and the time of a Greek doctor named Empedocles, who lived from *c.*484 to 424 BC. He theorized that there were four forms of matter (earth, air, fire and water) and that all matter was composed of 'infinitely small seeds' – what we would today refer to as atoms. It was Aristotle, however, who laid the foundations of science – and that of biology itself – by insisting that logic was the correct tool for investigation and deductive thought, and that observation and analysis were the essential prerequisites to study. A student of Plato, Aristotle lived from 384 to 322 BC in Greece and made a systematic attempt to organize all knowledge; in other words, to begin a system of classification.

Though his major involvement with biology was in the field of zoology, Aristotle began his classification of living things ('souls') by dividing them into three groups: vegetative, sensitive and rational. These approximated to plants, animals and humans, the only distinctions between the first two being made on the basis of movement, feeding and growth. Today's taxonomics call for more refined criteria, but Aristotle had taken that important first step. Indeed, there is evidence that he took the theory further and placed some animals in smaller groups, arranged according to common features. Aristotle's *genos* (meaning 'race' or 'kind') and *eidos* (meaning 'form') bore an equivalent status to our modern genus and species, though his definitions would be unworkable in today's classification framework.

Aristotle's influence was far reaching; it permeates much of what is accepted as commonplace today. Later, the seat of biological study swung from Greece to Rome, then to Arabian scholars, before finally settling back in Europe. Gaius Plinius Secundus, today known as Pliny the Elder (AD 23–79), produced an incredible *Natural History* some two years before his death in Pompeii, where he was attempting to witness the eruption of Mount Vesuvius at close quarters. The encyclopaedia contained detailed information on zoology as well as metallurgy, psychology and anthropology, but did not offer any improvement on Aristotle's method of classification. Even Leonardo da Vinci, in the last decades of the fifteenth century, failed to make any advance in this area and his bestiary included the sometimes wildly inaccurate details of animals he had found. The toad was included under the descriptive category of 'feeds on earth and always remains lean because it never satisfies itself, so great is its fear lest the supply of earth should fail'.

Early expeditions from Egypt and other 'civilized' parts of the world were usually made to bring back goods, and collecting plants or (less often) animals was no more than a sideline to these searches for useful products. Queen Hatshepsut of Egypt is known to have sent a specific expedition to locate a source of frankincense in about 1500 BC; the result was a selection of young trees brought back from Somalia, the first known botanical collection. Normally, in the forthcoming centuries, the drive was towards finding plants which could be used as food or medicine, and some species proliferated across continents as they were grown to order.

While these expeditions in search of useful plants were somewhat haphazard (there were few opportunities to exchange information between these early botanists), the drawings and notes they produced proved an exceptionally useful beginning to the documentation of plants. Many specimens were preserved dry, as there were no satisfactory methods of keeping them alive during transportation; this problem remained until the early 1800s when Nathaniel Ward invented a sealed glass container which retained its own stable climate: the Wardian case. The specimens in these dried collections, themselves hard to maintain due to problems of vermin and insect attack, were sometimes difficult to place in context with each other; one seed may easily look like another and, in any case, how could it be sufficiently well described in isolation, without the parent plant? For this reason, if for no other, attempts were made (some of them successful) to cultivate seeds. Lilies and bulbs were particularly successful, as they travelled well, and Turkey became a favourite area for collecting anemones, tulips and hyacinths. An outstanding example of a collector, scholar and subsequently author was Charles de l'Ecluse, who travelled widely and published a Spanish flora in 1576, describing some two hundred new species. Further works flowed from his pen, covering Austria and Hungary and then the whole of Europe.

Dried collections of plants and dead animals, either stuffed or with their mortal remains (horns, shells, bones) otherwise preserved, grew more popular through the sixteenth century. A number of individuals attempted to produce order in the collections which had arisen, relying on factors other than the use of the plant or animal. The result of this phase of collecting was, effectively, a series of private museums.

From the bestiaries of the Middle Ages come many of earth's mythical monsters, presented to the public as both fearsome and real. This drawing of a cetacean attacking a ship was published in 1555; illustrations of sea serpents and giant squid were common in the days before the great naturalists began their work.

The word museum is derived from the Greek *mouseion*, 'a site devoted to the Muses', or a place for learned discussion; it was certainly recognized that the contents of a museum were only of value if viewed in context with each other. The first natural history museum was probably that of a Swiss man, Conrad Gessner (1516–65), who used his collection as a tool towards the ideal of classification. Although the publication of his *Historia Animalium* in four volumes revolutionized the study of zoology and brought it from the times of Aristotle into the then modern world, he failed to discover any suitable underlying natural system of classification.

The 1600s brought a different approach. This was the embryo age of scientific advance: here were such eminent figures as Hooke, Halley and Sir Isaac Newton. Developed during the first half of the century, the microscope conferred a huge advance in the ability to observe detail. Newton gave us calculus and his *Principia* of 1687, in which the motions of the heavens were mathematically described. A rigid system of thought was present, and natural objects were classified into groups with features in common. Life was placed in categories, some resurrected from classical times, such as the animal, vegetable and mineral groups. In this, it was felt that all materials were alive: rocks were just slower growing than plants and animals, with a gradation in complexity. Evidence was pointed to: caves were old quarries, the passages becoming narrower as the rock grew, and how else would a stalactite form other than from a mineral seed? Fossils were encased in rock because that is where they grew and, by the late 1800s, there was also a theory that stalactites were a life-form midway between minerals and plants.

On the whole, the most practical system of biological classification remained with Aristotle and medical requirements: which herbs were important and how they could be identified. Classification was still according to use: could it make a tincture or ointment, was it poisonous or edible? Little advance had been made in producing order from disorder other than in this sphere.

For example, Dioscorides' *De Materia Medica*, written some time before AD 78, was an incredibly comprehensive medical text. Its drawings of useful plants were copied and copied again, and the volume remained in use for around 1,500 years. The *Ortus Sanitatis* of 1491 culled information on medically important plants, animals and minerals from all known sources and, along with other texts, became well known. However, the rationale of classification remained poor and, when the first English museum opened to the London public early in the seventeenth century, what became known as Tradescant's Ark was a simple collection of exhibits with little organization of the curiosities on show. Tradescant was a seasoned traveller, visiting North America three times and publishing a *Musaeum Tradescantianum* to document the ark's contents.

If the early and mid-seventeenth century was crucial to pure, physical sciences, its closing years and much of the eighteenth century was witness to surging advances in biology and classification. Both centuries were characterized by grand, exploratory expeditions across the oceans, which brought back the fruits (sometimes literally) of discovery. Drawings were made and a scientific approach of observing plants through all seasons was adopted whenever viable.

This process was usually one undertaken by men, but was not confined to them alone. When the Dutch lady Maria Sibylla Merian returned from three years in Surinam with a superb set of intricately observed illustrations of butterflies, caterpillars and their food plants, the resulting publication, *Metamorphosis Insectorum Surinamsum*, quickly sold out and was reprinted three times. Sir Hans Sloane (1660–1753) visited the Jamaicas and published a catalogue of his collection in 1696. Later the President of the Royal Society of London, a physician and a wealthy man, Sloane was able to spend up to £100,000 on his private museum, the equivalent of over £5 million today. When he died, Sloane gave his collection to the British nation, its 80,000 items forming the nucleus of what became the British Museum. Sloane himself was honoured when two parts of London took his name: Hans Crescent and Sloane Square.

At one time, the observable growth of stalactites and stalagmites was attributed to a mineral form of life. In fact, calcite cave formations are produced when carbon dioxide is lost to the cave's air from droplets of water, causing the calcite to precipitate and add to the structure.

Amateur science – especially in the realm of natural history – became a normal part of life for those wealthy enough to possess leisure time. For natural scientists a system of ordering the spoils of collecting was essential; such a tool was provided by Englishman John Ray (1627–1705), using anatomy as its basis. His three-volume *Historia Generalis Plantarum*, published towards the end of his life, recorded many of the then known plants, including those of North America. Ray was the first naturalist to introduce the modern concept of 'species': an organism which can breed with its own kind, but no other. Soon after, the Frenchman Joseph Tournefort (1656–1708) began using the concept of the genus as being a group of very similar species, and the stage was set for an essential breakthrough in classification.

On 23 May 1707, Carolus Linnaeus was born in Sweden. Son of a minister, he was trained in medicine but became keenly interested in botany in the 1730s; after all, botany and medicine went hand in hand. In 1741 he became a professor of medicine and botany, and through a stroke of genius developed an understandable, usable system of naming all living things.

The implication has been that, since Aristotle's time, organisms were only separated into plant and animal groups, or into herbal or food categories, and were not given any meaningful names. This was certainly not the case, but there was no consistency about the naming or organization of living things.

In an analogy, supermarket shelves might be stocked with food but, if products bear different names in different shops, there is no easy way to identify them or make an informed purchase. In the real world common names were in use, but different parts of the same country might call an organism by several names: what is termed a turnip in the south of England becomes a neep (a word based on the Latin for turnip: *napus*) in Scotland and the north, while a rowan tree becomes a mountain ash. A Reindeer (*Rangifer tarandus*) in Europe is

Horse-stingers, water-maidens, devil's darning-needles and goblin-insects are all outmoded names for dragonflies. The use of a common name creates problems in science; even today, damselflies are often incorrectly called dragonflies. It is only with the use of a Latin name that two people can be certain that they are talking about the same organism.

the same species as the Caribou in North America, while Europe's Elk is a North American Moose (*Alces alces*). To complete the confusing web, what an American calls an Elk (*Cervus canadensis*) is termed a Wapiti by a European. Introduce another language and the difficulties are extended still further. In 1885, discussing dragonflies, W.S. Dallas observed that what to the English were generally known as dragonflies were also termed horse-stingers (a misnomer, as dragonflies do not sting horses or anything else) and devil's darning-needles, while Germans saw them as *Wasserjungfern* (water-maidens) and the Swedes as *trollsländer* (goblin-insects), due to the appearance of the insect's head. Finally, across the mists of time, what we acknowledge within the sea as today being called an urchin would have confused our ancestors, to whom the word 'urchin' represented our modern-day hedgehog.

Science had also latched on to descriptive names, using Latin as the chosen, universal language. At least, the theory went, the writings of the Church were universally known in every country. The problem lay more with *how* to use Latin to describe living things, and scholars developed individual systems which ran to longer and more unwieldy names as all the prominent features were included. How many people would remember a name which included a plant's colour, size, leaf margin, number of petals, and so on, all translated into Latin? Frustrated by lengthy names, why not try size, colour or some other feature as the basis for all classification, as some scientists did. How about fruits, vegetables and nuts as formal groups?

The difficulties were rife; where would a new plant fit into the system? That a tomato and other tomato-like produce were vegetables was obvious, surely – yet, as they contain seeds, they are properly termed fruits. It was like trying to organize a stamp collection using the way a depicted face was turned, to the left or right, as the first, decisive feature for classification. The system would work to a point, but was hardly convenient; nor did it provide the most useful results for further study. While some schemes for naming organisms were developed and used, notably in private collections, they lacked any logical, extendible rationale and proved to be insufficiently flexible. Incapable of yielding useful information, they were unsuitable for universal acceptance.

Linnaeus attacked the problem in two ways. Firstly, he adopted what became known as binomial nomenclature (a name consisting of two words: the genus and the species) to give an individual organism a simple, unique name, and secondly he devised a classification system which would encompass all that was known in nature.

All matter had already been divided into three major kingdoms: animal, vegetable and mineral. As a botanist, Linnaeus worked with plants (the vegetables) and used the number of sexual parts (male stamens and female pistils, comprising the stigma, style and ovary) to divide them into groups which he termed 'classes' with their subdivisions of orders, then genus and species. At each descending stage the plant groups had more and more in common, in the same way that a series of filing cabinets might be used to subdivide information in an office: to find the right file, locate the correct room (kingdom), then the right cabinet (class), the drawer (order), the file within it (genus), and finally the correct sheet of paper (species). For example the Common Poppy, *Papaver rhoeas*, is grouped with other poppies which are similar in that they all have four petals, two sepals, and so on. The generic name of *Papaver* is shared with the other poppies in the 'file', like a surname. Each type of poppy also has a specific, or species, name (the forename, or given name) that is unique, in this case *rhoeas*. In contrast, the Opium Poppy is named *Papaver somniferum*, indicating something of its soporific effects.

Linnaeus published his work on plants in the *Species Plantarum* in 1753, following it with the tenth edition of his *Systema Naturae* in 1758. The latter, concentrating on the animal kingdom, was first published in 1735, but this later version had more than two thousand pages detailing six classes of animals. These comprised the mammals, birds, amphibians, fish, insects and worms.

Rather than only including what we think of as worms, the last category actually included everything that did not fit into the other, more obvious groups. Decisions as to classification were based on a simple key. A key is like a chart indicating features; for the first step, Linnaeus determined which group an organism belonged to on the basis of the number of chambers found in its heart. While other external features were also used, this meant that only dead specimens could be classified, and no distinction could be made between reptiles and amphibians as these have similar hearts. Nevertheless, more than 4,000 animals were defined, including humans which, for the first time, received a Latin name: *Homo sapiens* (wise, or thinking man).

Linnaeus was able to define an even greater number of plant species than animal species; on the basis of the number of stamens present, he divided some 7,000 plants into 23 classes, plus another class for non-flowering plants. The number of pistils then determined the order, and so on. It was a logical, understandable system which was immediately usable by everyone who could count, and purported to have revealed the underlying scheme established during the Creation. In 1761 Linnaeus was knighted by the Swedish government and henceforth used the name of Carl von Linné, rather than the Latinized version by which he is better known today.

The Foxglove, *Digitalis purpurea*, was the first source of the drug digitalis, used in heart treatments. The Latin name given to the plant by Linnaeus was a descriptive one, based on the finger-like shape of the flowers.

Although Linnaeus had proposed an excellent system, his work still had to gain acceptance outside Sweden. Georg Dionysius Ehret, a German illustrator, settled in England and promoted Linnaeus's work. Although it was immediately understandable and usable by everyone, it was nevertheless slow to come to full acceptance. The Church held sway over many aspects of science; in the mid-1700s purity of thought and a belief in the literal word of the Bible were all important. In this era what had Linnaeus done? He had constructed a universal classification system which depended on sexual parts of animals and plants for basic identifications.

Further, this was an age of the prim and proper. Not content with establishing reproductive anatomy as of primary importance in classification, Linnaeus referred to flowers with many stigmas and a single pistil as being 'many males in one bed with a female' (the Polyandria) or the rather more acceptable 'one male with a female' (Monandria). Petals formed a bridal bed. While scholars agreed with and applauded his reasoning, there was distress that such concepts would have to be dealt with by women and 'virtuous students'. By way of explanation, it was pointed out by aged naturalists that not every female or youngster would recognize the characteristic nature of the *Clitoria* genus.

However, most names were descriptive in other ways (such as the shape of a flower likened to a gloved finger: *Digitalis*, a genus including the foxglove), or were based on the classics or names already in existence. The principle of naming organisms after their discoverer, or otherwise honouring botanical men, was also established: the genus *Tradescantia* (a popular pot plant with variegated leaves) was named after John Tradescant, for example. In that age of scientific exploration, if a species could not be identified using Linnaeus's system it 'must' be new to science (with obvious pitfalls to such reasoning). Within only a few years, the system had been reprinted and appeared in any number of handbooks on natural history. In this field, it was a revolutionary concept.

The Linnaean system triumphed and demonstrably stood the test of time, as it is still in use today. As Linnaeus expected, there were later modifications, for example the addition of the intermediate groups of phylum and family: the sequence is now kingdom, phylum, class, order, family, genus, species. Nevertheless, even given these changes, his 1753 and 1758 publications on plants and animals are still regarded as today's baseline study from which all else follows, even after almost two and a half centuries.

Kindled with the ease by which plants could be classified, botanical (and, to a lesser extent, animal) collecting continued apace. In Africa, in the wake of missionaries, explorers returned with rich baskets of fare (the high point came when William Birchall collected over 40,000 specimens). China proved more difficult as, though plant sciences were relatively well advanced, foreigners were met with distrust. However, Russian studies were successful, as were major expeditions to the new worlds of the southern hemisphere.

Collections prospered and grew. In 1768 Captain James Cook sailed from Plymouth in the *Endeavour* with instructions to make astronomical observations at Tahiti and to search for *terra australis incognita*, the 'unknown southern land'. On board ship was an astronomer, artists to record whatever the expedition might reveal, and two botanists: Joseph Banks and Daniel Solander. With them went a copy of Linnaeus's twelfth edition of the *Systema*

The Duck-billed Platypus, *Ornithorhynchus anatinus*, was first brought to the attention of scientists at the British Museum in 1799. It was immediately branded as a fake as it possessed a bird-like bill (the genus name of *Ornithorhynchus* signifies this characteristic), duck-like swimming feet and the hairy, warm-blooded body of a mammal; it produced milk for its young, but had the egg-laying ability of a reptile. It was more than eighty years before the platypus was accepted as an egg-laying mammal, and classified as a monotreme which probably has its ancestors in Jurassic times. Today we recognize the difficulties of classifying new species, and know the reasons for at least some of the platypus's strange characteristics: the webbed feet are used for swimming, while the bill carries electrosensors (similar to those of the shark) to detect prey.

Naturae, the better to identify and catalogue species *in situ*.

This, perhaps the greatest of expeditions, returned triumphant on 13 July 1771. Apart from examining Tahiti for three months, rediscovering New Zealand, and surveying some 3,200 km of coastline along the eastern shores of Australia, it brought back a wealth of material and knowledge under Banks's direction. The die had been cast: to carry a naturalist during exploratory voyages was henceforth almost obligatory.

Banks had an immense influence on the way that collections were handled and organized. Born in 1743, he was President of the Royal Society of London for 42 years and wealthy enough to have helped finance the 1768 expedition. Indeed, that had not been his first exploratory collection, for in 1766 he visited Newfoundland and Labrador. The specimens he brought back formed what became the Banks Herbarium. Cook's South Seas expedition yielded over 800 new species (hence the naming of Botany Bay, the place of first landfall) and, when he was director of the botanical gardens at Kew, Sir Joseph Banks extended his influence to others by means of teaching; Robert Brown, for example, brought back over 4,000 specimens from Australia following an expedition from 1801 to 1805.

The trend continued. In 1799 a German, Alexander von Humboldt, sailed from Marseilles, in France, to return laden with the spoils of a scientific collection in South America in 1804. The botanist on the voyage was Aimé Bonpland; over 60,000 specimens of plants were collected, as well as a wealth of data covering the weather, fauna and geology. Humboldt's discoveries were published in his *Kosmos*, an attempt to set out the elements of scientific disciplines and make order from the universe.

Linnaeus had defined a system for naming and identification without being forced to resort to numerous, separate texts which themselves were inaccurate and difficult to use. His work was much revered and, following his death, Linnaeus's own collection was purchased and taken to England in 1784 (it was held in such esteem by the Swedes that, supposedly, a Swedish war boat followed in pursuit of the exported hoard), where it formed the kernel of the Linnaean Society of London collection. However, little new work was done in the advancement of Linnaeus's system until the early 1800s. At that time Baron Georges Cuvier, a French cabinet minister, became known for his classification of fossils. Cuvier's work enabled him to classify organisms on the basis of their anatomy and he extended Linnaeus's binomial nomenclature; published in 1817, Cuvier's *Le Regne Animal* was the first important advance for over fifty years.

With Linnaeus, Cuvier believed (as did virtually all other scientists, under the teachings of the Church) that all species were immutable; they were fixed, unchangeable and therefore, once classified, there was no further point in study. The drive was now towards classification for its own sake, rather than for use as a tool to further investigations. New species were being discovered with a rapidity which has been unmatched since those days, but much of the collecting was carried out indiscriminately in the hope of finding new species. In this arena, individual variation was ignored: a species, by definition, was unique. All differing specimens must, by implication, be new or, at least, not be observed too closely if the Linnaean system had already identified them. If nothing else, the concept of the type specimen was established, where the first formally identified specimen was preserved and (in theory at least) all other specimens thought to be of that type could henceforth be compared with it.

To a degree it might be imagined that scientific research was closeted away, that it belonged to those few who were wealthy enough to promote its ideals. Nothing could be further from the truth, at least in the loosest definition of nature study. The latter half of the eighteenth century (and, with gathering pace, the first half of the nineteenth century) was characterized – biologically speaking – by a need for legitimate entertainment which encompassed healthy exercise and religion. The answer lay with the natural world.

As affluence brought servants, and nannies for children, collecting became a leisure activity for the middle classes. A Sunday constitutional might be made in search of new specimens, rather than simply for exercise. Men, at least the few who could afford it, could legitimately shoot large game in the name of science as well as sport. A new species, after all, could be named for its discoverer, a high point in any career and a chance for immortality. Beetles and birds, flies and fish, all became grist for the collector's cabinet, and many people maintained a separate selection of 'rarities' which therefore, presumably, became rarer still. In 1789 the Hampshire vicar Gilbert White, more sedentary than men like Banks, produced *The Natural History and Antiquities of Selbourne*, in which observations were laid out in a series of letters. It was one of the few examples of a home-grown nature study which depended on fieldwork, and proved a huge stimulation to those of like mind: it was *not* absolutely necessary to travel far afield to maintain a collection.

From a position of paternal tolerance by society, collecting became an obsession which shifted from insects to shells, to seaside rock pools and back again via ferns and legendary beasts. If an object could not be taken back to the drawing-room, then drawings themselves sufficed. If myths prevailed, such as the oft-repeated fact that frogs could live sealed within a rock (experiments were undertaken at Oxford to determine the 'truth'), then it added to the wonders of popular science. There were local crazes: limpet collecting or keeping pet alligators. If a collector or illustrator required a new, impressive subject, there was always the art of invention, something that the collector Rafinesque was renowned for.

For a while in the early 1800s, Rafinesque held the position of Professor of Natural History at a university in Kentucky, a title which lent him credence. However, the French–American ornithologist who gave his name to one of the best-known American

A dragon 'from the Caverns of Mount Pilatus', which guarded the riches of the earth. Taken from the Revd Kircher's book *Mundus Subterraneous*, first published in 1665, this dragon myth may have arisen from the presence of *Proteus anguinus* washed into the River Lintvern (*lintvern* translates as 'dragon') from caves near Trieste.

Proteus species are sightless, colourless, cave-dwelling salamanders. Kircher, interestingly, pre-dated two later theories when he argued that diseases were transmitted via 'contagium animalium' (the equivalent of today's bacteria) and that 'modern' species developed by transmutation from previous life-forms.

societies, John James Audubon, played pranks on Rafinesque by describing fictional species. Audubon's jokes did nothing to promote the natural sciences but, at the same time, the public seemed to care little for rational arguments as to the existence or otherwise of this or that monster; to do so might rule out an exciting possibility. Why deny the report of a dragon when such beasts *might* exist? The excitement engendered by journeys to foreign lands continued, such as Charles Waterton and H.W. Bates's separate expeditions to South America and Paul du Chaillu's to equatorial Africa in the 1850s and '60s. Rich pickings by way of specimens were available for the taking and a fascinated public ensured a financially satisfying return for the more obscure and sensational, such as du Chaillu's gorilla remains.

By the early 1800s natural history was seen as the study of what are now the three sciences of biology, chemistry and physics, plus mineralogy and geology. To be a naturalist indicated an interest in all these aspects, though the fields of study were slowly separating, with the term 'biologist' being introduced in the middle of the century to separate the professional scholar from the amateur. Today, to be termed an amateur is frequently but not always accurately taken to mean 'less good' at whatever endeavour is being cited, even though the very word amateur indicates a 'love of the subject'. Then, it was more a term of class distinction: only those wealthy enough to support their obsession could become a professional scientist, as most appointed posts paid little.

To much of the population the 'science' of nature study amounted to an opportunity. The first half of the nineteenth century saw numerous British field clubs and societies spring forth. Ladies could join and, indeed, might even be welcomed for their skills in collecting (although it was felt that only men had the ability to classify correctly and thus identify new species). In fact, some field meetings were little more than a gathering of like-minded people in search of relaxation, and might be linked to a picnic or outdoor lecture. A great deal of effort was expended in collecting specimens and, in preference to reading, cataloguing the day's finds. For these people nature offered the semblance of joining with the great naturalists of the age and tasting the excitement of this new scientific era.

At that time classification was still the objective of science; new theories were rare and generally shunned. Within Church doctrine, to study with no purpose was blind curiosity and should be condemned, while study with purpose – such as towards medicine, or classification – was praiseworthy. Collect, collect, collect was the order of the day. If it could not be collected, then measure it: measure rat tails and pig tusks, collect and quantify and collect some more. There was an obsession with keeping checklists and ticking off new finds, obtaining sets of organisms, and being methodical and orderly. Now that plenty of curiosities were being brought home and a neat, understandable system of classification remained in universal use, all that was required was to fill in the gaps. Sir Ernest Rutherford, the physicist, later called these naturalists 'glorified stamp collectors'.

In this age of industry the study of nature was seen as bringing people closer to God, as it was a study of what He had made. Intricacies discovered through the microscope indicated His existence as, of course, mankind could never hope to challenge this minute art. Popular texts on nature began to appear, encompassing the wonders of nature and how they indicated the marvels of Creation. In 1866 an anonymous author wrote:

> *The earth, the sky, the sea, and the 'waters under the earth,' abound with marvels calculated to excite astonishment and profound veneration for the Divine agency through which they have severally been wrought. The tiniest blade of grass, the smallest imaginable drop of water, the minutest particle of impalpable air, is to the reflective mind an object of special wonderment, offering a wide field for patient investigation. But how much more deeply is human admiration stirred by the contemplation of the stupendous works with which the Almighty has signified His powers and His designs!*

Natural history study nevertheless expanded its influence, with the blessings of the Church. There were even attempts to explain how the classification of animals and plants might fit into the overall scheme of things without challenging Genesis. That God had created all things as immutable was a basic, underlying fact that, to the Church, must be accepted without question and, by and large, naturalists saw no difficulty in operating within this framework. Nevertheless, there were anomalies.

How had fossils been produced? Cuvier believed, as he should according to accepted doctrine, that all species had been created together, but he felt that some had later died due to natural events such as volcanic eruptions or floods (prior to Noah's deluge, for that was the time when all living things were saved). There were even attempts to explain the distribution of fossils by suggesting successive catastrophes, each wiping out organisms which were then replaced by God (the theory finding acceptance with the Church as there was no conflict between Genesis and science).

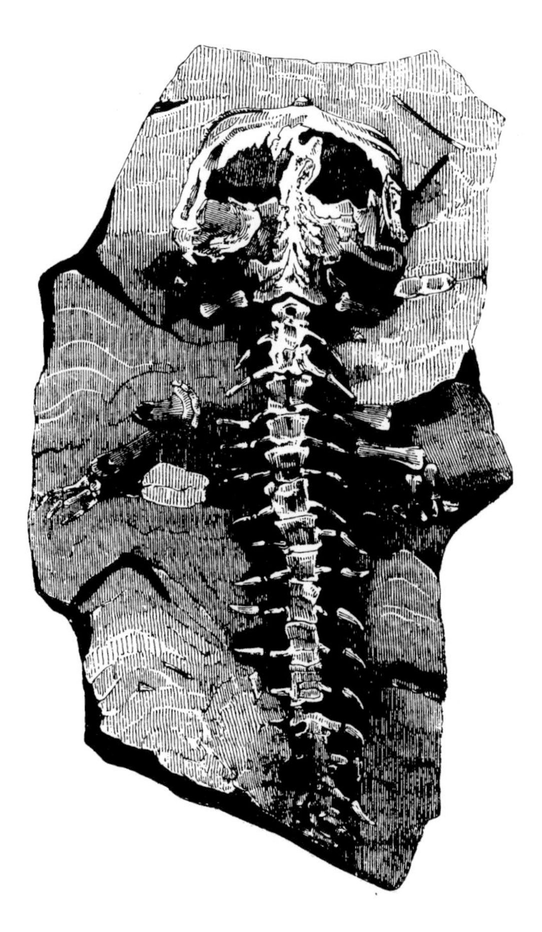

The supposed fossil of a man, discovered in 1726. It was thought to be antediluvian – that is, it pre-dated Noah's flood and represented one of the 'cursed race' wiped out by an act of God. In later years the fossil was shown to be part of a salamander.

Around the same time, 1809, Cuvier's fellow Frenchman Jean Baptiste Lamarck postulated that plants and animals could change according to their environment and desires. A man who developed muscles during his manual labour would beget sons who were similarly strong, while short-necked giraffes straining to reach leaves on high branches begat young which possessed longer necks because their parents desired it. The theory of Lamarckism indicated that such changes could be passed on to the next generation.

The theory offered a false road which was nevertheless taken up by several authorities, notably in the USSR; Cuvier did his best to crush the idea from existence as being unacceptable in the light of his belief that all things were immutable, the basis of Church-influenced classification. While Lamarck's theory was discredited (and remains so today, though elements of his teachings have been incorporated into other works) Lamarck did extend Linnaeus's work on animals, and Cuvier's studies, when he distinguished between animals with backbones and those without, coining the names vertebrate and invertebrate and, incidentally, using the term biology for the first time. Linnaeus's mish-mash classification of virtually all invertebrates other than insects was also subdivided into arachnids, crustaceans and echinoderms, plus what still remained, such as worms. In addition, Lamarck introduced a system of identification which used a series of either-or questions to lead to the name or group of an unknown organism. This concept of division and subdivision produced what is termed a key. For example, to classify which of today's group of animals an elephant belongs to, a key might ask:

1. Does the animal possess a backbone?

2. If so, is its skin scaly and wet, smooth, scaly and dry, feathered, or hairy?

The answer to the first question would identify the elephant as a vertebrate. The skin features in the second question would then reveal the

vertebrate to be a fish, amphibian, reptile, bird or – as applied to the elephant – a mammal. By following the same procedure, questions could lead through all the characteristics of the organism to reveal its Latin name.

Of all the theories which affect our modern-day biological science, none are better known than those of naturalist Charles Darwin. He successfully challenged Church doctrine in respect to Genesis, though this was not his intent, and developed a theory to explain how variations could lead to a selection process, which in turn would produce new species. Even so, the theory had a long gestation and was not the only one extant. As with nearly all major advances, it was built upon the work of others and brought to fruition by circumstance and a touch of genius.

The African Elephant, *Loxodonta africana,* is defined as a mammal as it has a hairy skin, as well as bearing live young which are suckled with milk; the group is named after *mamma,* the Latin word for breast. Mammals are also warm-blooded and possess a spinal column.

Darwin was born in the English town of Shrewsbury in 1809, the son of a physician. His upbringing was normal for a child of his background, in that science was not taught at school and scientific facts were believed by some to be unfruitful in teaching. Experimentation was regarded as a further waste of time and Darwin was prevented from following any such course.

Initially undertaking training in the medical profession (the sight of unanaesthetized operations being performed finished *that* course of study), Darwin was instructed in the holy orders at Cambridge, while at the same time becoming more and more interested in natural history. When he was offered the chance of taking part in a five-year voyage to the Pacific as a naturalist, he took the opportunity and, at the age of 22, set sail in December 1831 on HMS *Beagle*.

Far from being a genius with new ideas at that time, Darwin was a product of his age. His grandfather, Erasmus Darwin, had formed some embryonic ideas concerning inherent changes in populations, which Darwin rejected along with Lamarck's theories; a religious training would have directed such thoughts. However, during the voyage, a copy of Lyell's *Principles of Geology* caused him to think further. Lyell believed that geology was subject to continuing forces which caused changes in the rocks around him. What if changing forces were brought to bear on living organisms?

Some of the specimens that Darwin collected, and his observations, puzzled him. He observed that changes in coral reefs must occur over time, as new reefs were built upon the framework of older, dead structures. To explain this required a radical thought: that ocean levels were not constant, or that land surfaces must sink or rise (Darwin found fossil seashells high in the Andean mountains). As this occurred, coral would grow to accommodate the change in the environment. Nature was not immutable.

Further, he found that species in different geographical areas might be similar to each other without being identical, the classic example being finches on the Galapagos Islands. Here, the bill shapes of the birds were different from those discovered on the mainland of South America, and Darwin believed these were ideally suited to the birds' differing diets. Even the islands yielded slightly different populations, depending on the food available; a large, strong bill suited a diet of large seeds, slender bills with comparatively little strength were ideal for catching insects, and so on.

Charles Darwin at Down House.

Though this example is probably the most famous, it was almost chance which led Darwin to note the differences between islands. The vice-governor remarked to Darwin that it was possible to tell which island was which given sight of its tortoises. The idea that each island's fauna differed slightly caused a flurry of activity, Darwin rechecking his data (much of which was only recorded as being 'Galapagos Islands' with no distinction made between them) within a few hours of sailing. In fact, it was Darwin's minutely detailed notes on plants, combined with fossil and other evidence, that clinched the new-born theory. Darwin concluded that successive generations of a species would slowly alter their characteristics to suit their environment; slight differences between islands were enough to accentuate slight differences in individuals until new species were formed. This differed from Lamarck's theory in that it did not depend on acquired characteristics being passed on to offspring, but on variation between individuals being sufficiently powerful to account for changes through time.

Darwin hesitated to present his theories, which he developed over subsequent years, until he had gathered more evidence and worked further upon his specimens. Theories were not always popular (look what had happened to Lamarck and others) and Darwin wished to back his with scientifically gathered facts. In particular, while he felt that changes to species could occur, he needed to find the mechanism which enabled this. In 1838 he read and was further influenced by theories of population growth propounded by the economist Thomas Malthus. In 1798 Malthus had suggested that human populations, growing in the wake of the Industrial Revolution in Britain, were not necessarily changing for the better. If human populations were altering in the face of environmental change, what if the same applied to animals and plants?

The time was certainly over-ripe for such thoughts. As far back as the fourth century BC Empedocles had postulated, in a poem, that organisms were changeable under naturally occurring forces. In 1813 William Wells presented a paper to the Royal Society of London in which he discussed the origin of the Negro race and, further, suggested that there was a selection process at work which depended on variation within a species. The ideas were well in advance of that time, but the paper was ignored and seemingly unknown to Darwin. In 1842 Darwin wrote what was effectively a preliminary paper on his 'new' theory, and a longer essay in 1844, but they were known to only a few men and not widely disseminated or formally published.

In the same year, 1844, Robert Chambers published *Vestiges of the Natural History of Creation*. In this, Chambers expressed the view that animals and plants had changed over the millennia. However, it also drew on tenuous evidence and called for the belief that man had descended from a large frog. Naturalists of eminence, including Darwin, denounced it, but there were yet other publications hinting that Genesis might not be scientifically correct. Was there really a monster of the deep (Leviathan) or of the land (Behemoth) as the Book of Job recorded, and, if they did exist, where were they now? Of more importance was the fact that a number of theories intended to tackle the problems of 'transmutation of species' were at last being discussed.

Finally, in 1856, Darwin began to write an expanded version of his theory of natural selection, which indicated – still in the face of accepted teaching – that organisms were malleable. But, in 1858, before it could be completed, he was contacted by Alfred Russel Wallace.

Wallace had taken part in expeditions to South America and Malaysia, and was also influenced by Malthus's paper. In 1855 he had already written a document, 'On the law which has regulated the introduction of new species', and suggested to Darwin that a natural process of selection was the mechanism for change in populations. The well-reasoned argument came as a complete surprise to Darwin, who was utterly dismayed that his own thoughts had been presented so well by another man. The outcome was that, jointly, they presented papers to the Linnaean Society in 1858. The following year Darwin's monumental work *On the Origin of Species by Means of Natural Selection, or the Preservation of Favoured Races in the Struggle for Life* was published. Darwin felt he was conclusively correct, as not only was the theory understandable and consistent, it was backed up by observation in artificial selection: he had spent years studying pigeon and dog breeding, producing 'new species' at an accelerated rate.

When Darwin's *Origin of Species* was revealed, a scientific furore spilled over into public life as the ideas attracted or repelled naturalists and common people alike, polarizing his peers. That there was intense interest after the papers had been read in 1858 is perhaps an understatement: the first printing of 1,250 copies of Darwin's book sold out in one day and, a few months later, it was followed by a further 3,000 copies; there were seven editions in his lifetime alone and the book has never been out of print.

In essence, Darwin's reasoning was that:

- Organisms produce many more offspring than can possibly survive; if they did survive, the earth would quickly be covered in mice, or elephants, or frogs. If mice can produce ten litters of five young mice per year, and each of those survives to breed, in one year there would be 4,500 mice; a single batch of cod's eggs, meanwhile, would result in the birth of 1 million progeny. Darwin estimated that, even with a long gestation period of 22 months, a pair of elephants would be capable of yielding 19 million descendants within 750 years.
- Such a result, due to unlimited reproduction, with every individual surviving, is an obvious nonsense. Darwin deduced that the vast majority of seeds or young must die before reaching maturity, in order to account for the fact that populations remain relatively constant.

As Darwin noted, far more young are produced in the wild than could ever survive without quickly over-running the earth. His conclusion was that there is a selection process in force: only the best suited to the environment grow to adulthood and breed. As an example of the degree of excess production, this British Common Frog (*Rana temporaria*) may lay up to 2,000 or more eggs each year. Other species, such as the American Bullfrog (*Rana catesbeiana*) and European green frogs, can lay over 10,000 eggs.

- A study of any species based on its reproductive parts and anatomy, in other words using Linnaeus's system, indicated that, while individuals from the population were undoubtedly the same species, there was a wide range of variation in their characteristics. Some individuals might be small while others were large, or possessed different colours, a greater wingspan, and so on.

- If the majority of a population was destined to die before maturity, and individuals varied one from another, only the best suited of these different individuals could be expected to survive, reproduce and pass their characteristics to the next generation. The weakest would die young, and the strongest survive longer. Thus, the concept of the survival of the fittest was born. The variations best suited to the environment conferred an advantage for survival.

To illustrate the point, take the following example. In times of food shortage the animal which can run the fastest might have the best chance of survival in catching prey, or of escaping. On the other hand, the same set of animals might find that, in a time of plentiful food (when this factor is therefore not so important), an individual which is resistant to disease fares better in the survival stakes. The variations in the population are random, and what is of benefit to the individual depends on the environment. If an organism – plant or animal – bearing a particular characteristic is less likely to survive, the proportion of that characteristic present in the following generation is likely to be lower than it was before. The population has changed and, given enough generations, undesirable characteristics (for that particular environment) disappear. The same, original population, separated by some geographical boundary (such as mountains or an ocean separating islands from each other) and living in a different environment, might find other characteristics desirable. The end effect would be to produce new, different species from a common ancestor, given enough time.

With this explanation, Darwin had his mechanism. Wallace had also developed the same ideas and deserves far more credit than he is generally accorded, but Darwin's work was more imposing, backed with examples and discussion. Natural history had now been presented not only with the means of classifying all animals and plants in a logical manner, but also with the method which had produced that same myriad of species. But could it be accepted?

Darwin himself did little to defend his theories because he was not well and spent his life at Down House in rural Kent. It was left to the talented and respected Thomas Huxley to promote the idea of natural selection when *Origin* came to the attention of the clergy and other die-hard naturalists. It was Huxley, above all others, who presented the basis of what became known as evolution to successive audiences.

There were obvious difficulties, many of which have been widely reported. Darwin was ridiculed for daring to infer (he had not stated as such) that man was descended from apes; the inference was enough and, rather than referring to Darwin's 'transmutation' or 'descent' theory, reference to the 'ape theory' was more common. That others had indicated man's lineage years before was scarcely relevant: Linnaeus had placed man in the same genus as the orang-utan, for example. The 'monkey' aspects of the discussions were fuelled by du Chaillu's return from Africa in 1861, with the stuffed skins of gorillas and lurid descriptions of how they had been killed:

> *His fierce gloomy eyes glared at us; the short hair was rapidly agitated, and the wrinkled face seemed contorted with rage. He was like a very devil. . . . I waited, as the negro rule is, till the huge beast was within six yards of me; then, as he once more stopped to roar, delivered my fire, and brought him down on his face dead.*

In fact, there was some dissension between du Chaillu's account of dire battles and the bullet holes noted in the backs of the gorillas. While the evidence of their existence was not relevant to the theory of natural selection (in any case, specimens already existed in the USA), the animals were obviously man-like and fuelled the furore. It was not until the 1870s that the dissension died down. Even then, books on natural history either ignored Darwin or their authors found ever more contorted reasons to show that there was no dissent between his theories and that of the Creation. It was a backlash of 'natural theology' carried to extremes. Edward Hitchcock wrote a volume on *The Religion of Geology* in 1860, in which he discussed the argument that:

> *All the rocks and their contents were created just as we now meet them, in a moment of time; that the supposed remains of animals and plants, which many of them contain . . . were never real animals and plants, but only resemblances; and that the marks of fusion and of the wearing of water, exhibited by the rocks, are not to be taken as evidences that they have undergone such processes, but only that it has pleased God to give them that appearance.*

One explorer, Ribeyrolles, wrote after a visit to the forests of Brazil:

> *I admire those learned men who, bending over some herbal, bid you examine closely the structure of the internal tissues; notice the number of the absence of the cotyledons; follow the evolution of the seeds; verify the sexes; and then determine to which of the great classes of the vegetable kingdom it belongs.*
>
> *Is it really so easy? Is the life of plants simply a question of cotyledons?*
>
> *God forbid that I should denounce genius and patience. The great masters of botany have deserved well of mankind in laying down, as rules for investigation, certain natural affinities and organic analogies. They have set up the workshop, and simplified study. But how far have these classifications and methods contributed to reveal the* being, *the life, of a plant? Description is not explanation, and phenomena are not laws. Arrange as many museums as you please; construct cabinets and hothouses; go into virgin forest, and amuse yourself by counting cotyledons. What does it amount to?*

Some nineteenth-century 'collections' were little more than excuses to hunt big game, such as this gorilla propped up by native Africans. Du Chaillu's reports of savage attacks, in particular, were considered to be exaggerated, but the gorilla remains he brought back to Britain were nevertheless fuel to the Darwinian furore.

Taken from an immensely popular book published in 1866, *The Wonders and Beauties of Creation*, 'as portrayed by Humboldt, Livingstone, Ruskin and other great writers', the text was intended to impress the reader with not only the magnificence of nature, but also the excitement of expeditions to far-flung lands. Although it remained in print well into the next century, through the medium of several editions, neither Wallace or Darwin, nor any hint of their theories, received a mention. Darwin wrote that there was an 'almost universal belief' in evolution only nine years after publication of *Origin*; in this perhaps he was too hopeful. He died in 1882, to be accorded the honour of burial in Westminster Abbey in London.

Darwin's work, though crucial to the theme of classification and how species arise, is not without flaws. During his lifetime one recurring problem was a failure to provide sufficient 'missing links'; if one species could give rise to another, there should be intermediate, fossil steps preserved in the rocks. Some were indeed discovered, such as *Archaeopteryx*, which indicated the possible leap from reptile to bird. However, while variations could be used to show how new species might arise through natural selection, how did the variations themselves arise? Here, the laws of inheritance remained stubbornly unknown.

Victorian science correctly observed that if two parents had offspring, these would possess a mixture of characteristics from both parents. However, the theory then indicated that any one characteristic would therefore eventually be diluted to return the organism, over successive generations, to a basic form. This perfectly matched the idea of the type specimen and explained, in opposition to Darwin's theory, how species were immutable even though variations were observable. Darwin himself experimented with mice and plants to try to discover the mechanism for passing on characteristics but, though he was on the right lines, he failed to interpret his findings other than to postulate that freely-mixing 'gemmules' existed in all cells. These, he thought, could circulate throughout organisms and be passed on to offspring. These gemmules represented characteristics (there was even a hint that they could be acquired during life, a suggestion dangerously close to Lamarck's) and could lie dormant, only to reappear in successive generations. Experimentation failed to support the theory, for Darwin's cousin, Francis Galton, transfused blood from one rabbit to another but did not discover any change in the resulting young. The scientists of the day still required a mechanism to introduce the variations which natural selection worked upon.

In part, this was supplied by August Weismann. His brutal experiments, conducted in the 1860s and '70s, involved removing the tails from mice: no matter how many times this was done, the parents never produced tail-less offspring. It was the final nail in the coffin of Lamarckian theory. However, Weismann also noted a distinction between body cells and the 'germ plasm' of reproductive cells: what affected body cells would not necessarily affect sperm and ova, and hence the next generation.

Unknown to any of these biologists was the work of an Austrian monk, Gregor Mendel. Working in his monastery garden, Mendel bred pea plants and, in 1865, deduced that physical characteristics could be passed on to new generations. More, he concluded that these characteristics were not mixed in the way that paint might be, but remained distinct. If the choice was between white and grey fur, as with Darwin's mice, the result was invariably white, grey or piebald, but never a compromise of colours. Pea plant varieties which grew tall or short would always produce mixtures of tall and short plants when the seeds were planted, never an intermediate size.

Mendel presented his ideas to the Natural History Society of Brunn in 1866, where they lay forgotten until 1900. Then, his paper rediscovered, it was realized that here was an explanation – backed by scientific evidence – for passing on characteristics. In addition, it showed how these 'physical factors' (the term 'gene' was coined at a later date) could lie dormant to reappear after seemingly becoming lost to the species. The means for preserving variation was thus explained, though it was not accepted by most biologists until the 1930s.

Mendel's results, and successive experiments, did much to support Darwin's theory, but a way of producing *new* variants was required. It was noted that the physical particles which represented the characteristics (the genes) might be altered by artificial or natural causes: a mutation (the word means 'change') was possible. In 1953, when James Watson and Francis Crick showed that the chemical DNA was the basic element of genetic material, and that it permitted a genetic code to be copied during reproduction but could also be altered by chemical or spontaneous means, the question was answered. Evolutionary change could at last be explained in full, backed by evidence: species arose due to environmental pressure acting on naturally occurring variations, and new, permanent characteristics could be introduced by physical means. Animals and plants which adapted to a changing environment prospered, while those which could not – perhaps because they had become too precisely adapted to a particular niche within their habitat – died out.

This explanation of successive theories giving rise to yet more, in a chain of consequences, is rather too neat to accept without also realizing that alternative ideas and much dissent arose along the way. Many other concepts exist and theories such as Lamarck's still have their proponents. Modern ideas incorporate those of Darwin and Mendel, plus the knowledge of biochemistry gained by Watson and Crick. On the other side of the coin, some religious groups retain a basic belief in the letter of Genesis; it is sometimes difficult, as this short historical study demonstrates, to separate a belief based on faith from one based on fact when neither supports the other.

Because a corn cob bears numerous, separate flowers, each is individually fertilized and produces kernels based on different genetic combinations. In this example, 'genetic corn' has been produced by controlling its pollination; the result is corn which exhibits kernels with different characteristics, most obviously that of colour.

In particular, the interaction and prevalence of genes is indicated by the proportion of different kernels in successive generations. Mendel used similar results obtained from peas, using the appearance of the seed and the size of the plant to determine his laws of genetics. Examples such as this can indicate which genes dominate over others and their relative proportions in the specimen.

It is also true that, as with all theories which underlie other research, attempts to make evolutionary theory too simple, of popularizing the concept, are a danger and can lead to pitfalls. It should also be borne in mind that evolution is not under some careful, directed plan: the production of variations, upon which evolution ultimately depends, relies upon a random process, in just the same way that a coin will randomly yield heads or tails. Add the possibilities encompassed by a larger pool of options, such as rolling dice or cutting a pack of cards, and the random nature of the process becomes even clearer. When a new variation springs up it might prove advantageous, of no benefit and therefore neutral, or cause a disadvantage to the organism which leads to its death or an impaired reproductive potential. Pure chance has, according to Darwinian theory, provided us with our teeming earth of fascinating species, which can be ordered according to Linnaeus's rules.

For many years natural history preferred to remain with the cosy feeling that everything could be easily and readily categorized; it was only a matter of time before 'missing' species were neatly placed in their correct position in the great tables. Eventually, while Linnaeus's work was crucial and still underlies all modern biological study, it was the impetus that Darwin invoked with his carefully reasoned ideas that forced biologists from their dry rooms and into the field with a new aim: rather than cataloguing dead organisms alone, the theme became one of studying life.

CHAPTER TWO

IT'S LIFE, JIM

And God said, 'Let the earth put forth vegetation, plants yielding seed, and fruit trees bearing fruit in which is their seed, each according to its kind, upon the earth.' And it was so.

And God said, 'Let the waters bring forth swarms of living creatures, and let birds fly above the earth across the firmament of the heavens.'

And God saw that it was good.

Genesis 1:20–1

It may be a matter of little importance to us at what precise moment the first living creature saw the light of this world; others may speculate as to whether that first dawn of vegetable and animal life brightened millions or myriads of years ago, whether our Creator imparted life to lifeless matter by a word, or by the slow, silent action of cosmical forces: enough that at a moment unknown to us life sprang into being upon our earth.

Klein & Thomé, *Land, Sea and Sky*, 1881

WHEN NASA BEGAN sampling the surface of the moon, returning cargoes of dust and rocks, there was the faint, faint hope that traces of life might be found. The possibility was taken seriously after the first manned landing on the moon in 1969; on their return the Apollo 11 astronauts were placed in 'biological isolation suits' before leaving their capsule, and quarantined for three weeks. Within the framework of *Star Trek*'s medical section genius, Dr 'Bones' McCoy, anything they found might be 'life, Jim, but not as we know it'; nevertheless, it would have been some recognizable form of life: it would be classifiable.

If life had been found on the moon, NASA's hypothesis was that it would have to be extremely resistant to cold, dehydration and radiation. Such factors are not unknown, even on earth. Tardigrades (small, fascinating creatures commonly called water bears) can withstand total desiccation and freezing in Arctic tundra for years on end, and a bacterium, *Deinococcus radioduran*, has the ability to manipulate and repair its DNA (deoxyribonucleic acid) after exposure to levels of radiation several thousand times higher than that required to kill any other known species. However, despite science fiction's claims that the weird and wonderful must inhabit every turn of interplanetary travel, NASA's examination showed the moon to be barren and dead. If there is life elsewhere in this expanding universe, such new kingdoms must await the biologists of the future. Here, on earth, there is enough to do: the classification of life has barely begun.

This does raise a question, though. How do we recognize any life-form as living in the first place? From its simplest to most complex, what is a definition of life – and where did it come from?

In response to Darwin's theories, in 1871 the President of the British Association, physicist Sir William Thompson (later Lord Kelvin), publicly announced that life had been spawned on earth as the result of 'countless seed-bearing meteoric stones'. His evidence, as might be imagined, was slight. He also deduced the age of the earth, extrapolating back using the cumulated heat loss of its rocks (as had Buffon in the eighteenth century); his figure of 98 million years old was at least more accurate than that of James Ussher, Archbishop of Armagh.

In the mid-seventeenth century Ussher deduced, using biblical genealogies, that the earth had come into being only 6,000 years before; this was later refined to Sunday 23 October 404 BC. On the basis of then current beliefs it was a reasonable piece of research and was even printed in the margins of bibles. Later, the date proved problematical when geology and fossils suggested a much older age for earth, until Cuvier suggested that there had been several Creations, each replacing the previous one after successive floods and that the rocks were indeed older than the latest Creation. Such ideas did much to restrain advancing theories of evolution.

The geologic history of the earth (we now believe that the earth is about 4,600 million years old) is divided into four segments, the earliest being the Precambrian. This, alone, accounts for some 85 per cent of geologic time, and was followed by the Palaeozoic and Mesozoic eras before reaching the Cenozoic. This era has existed for 65 million years, to the present day. If the numbers are unimaginable, consider the distance you could cover in one hundred paces. If that would take you to the dawn of time, the dinosaurs died out less than two footsteps away, and man appeared on the scene within the space occupied by a small toe.

Life is known to have existed from at least the early Palaeozoic, as fossils in rocks of this age suddenly appear in abundance. Living organisms must have developed prior to this time, however, for these fossils show complex designs which, evolutionary theory tells us, must have arisen from basic forms. Evidence of earlier, simple algal cells exists in the shape of stromatolites, limestone structures which developed when algae trapped sediments, and in the fossil remnants of worm-like Ediacaran animals (named after a range of Australian hills) from some 600 million years ago, although we believe that the first glimmerings of living cells arose even earlier, far back into the Precambrian. That fossil trilobites and molluscs are found in such numbers is due to the presence of exoskeletons and shells, rather than the soft bodies of earlier organisms.

But how do we recognize these organisms as having once being alive? What is life?

Ask a child if something is alive and there will usually be an instinctive answer to a pointing finger: yes or no. How does he or she know? Let us suppose the classic cartoon were to come true: a 'beep-beep, bug-eyed, green-skinned Martian' lands on earth, walks up to the nearest petrol pump and states, 'Take me to your leader!' We laugh: hasn't this Martian got any intelligence? Why should we assume it can determine the degree of life present in this artifact, which it has never seen before?

When it comes to defining life, schoolroom teaching uses a variety of characteristics as indicators. If, the definition goes, an organism exhibits all these characteristics then it must be alive. Thus, a cat and a rose tree are both alive as they both grow, require nutrition, excrete, respire, move, are sensitive and can respond to their surroundings, and can reproduce. There is no anomaly between using these seven characteristics of life for both plants and animals; the method of nutrition and movement may be different in each group, but 'feed' and move they do. Likewise, the degree of sensitivity and speed of reaction might make plants appear radically different from animals (plants can detect light and gravity and orientate their growth accordingly), but the same criteria are in force.

So, the petrol pump (or brick, or car, or building) is not alive because it does not fulfil all seven criteria. What of the car? Surely, it is alive: it can obviously move, it respires (takes in oxygen), requires nutrition (petroleum), excretes (expels exhaust gases), but while you

Are these flowers and the butterfly alive? Of course: each grows, feeds, moves, reproduces, and possesses all the 'seven characteristics of life'. If this is our definition, though, can something which is not capable of reproducing – such as a virus – be considered alive?

might argue for sensitivity (it changes direction when a wheel is turned) it most definitely does not grow or reproduce. So, non-living things may possess *some* of the characteristics of life, but not all.

Yet, this form of definition does not perfectly match all our conceptions of life. A cat or a tree are easy fare, but what of a virus?

A virus is a form of microbe (a microscopic organism) which infects living cells. It may be one of several shapes or sizes, the simplest being a rod, but it is many times smaller than a bacterium. To use the term germ instead of bacterium would be a misnomer, as not all bacteria cause diseases. However, a virus, by its nature, always causes damage. Indeed, it was through the study of viral infections of bacteria that much of our detailed knowledge of them was first obtained. The interesting point is that a virus is incapable of reproducing itself without using material from other cells; it has no inherent, internal capability of its own. A virus relies upon the contents of the cell it is attacking to produce copies of itself which then spread to other cells, in so doing sometimes carrying pieces of the host's genetic code to another individual. This is a process termed transduction, which can unwittingly introduce new variations to a species. The fact is: outside a host cell a virus is no more than an inert, non-functional chemical.

So is a virus living? This depends on your definition: it is not alive under schoolroom criteria, as it does not contain within itself the means to reproduce, and in addition it probably does not respire. To be able to argue for viral life, the world of chemistry must be consulted.

For all that life on earth is stunningly diverse, the chemicals which form it are based on a single element: carbon. As an interesting sideline, the most popular science fiction alternative is silicone as this bears the required structure to bond readily with other 'life-giving' elements. However, under 'normal' temperatures and pressures silicone does not have the capability of forming the complex compounds that carbon can, though it does possess the same four-bond ability to join with other chemicals. Science fiction must look to very different planetary conditions to hypothesize the existence of a silicone-based life-form, though it is nevertheless the material that we use for computer chips in our attempt to develop artificial intelligence. Be that as it may, carbon is a common factor to all known living things.

Carbon is such an important element because it combines with other elements such as oxygen and hydrogen to form large molecules, in particular nucleic acids and proteins. Processes within all cells depend on complex proteins called enzymes, which alter the speed of chemical reactions, usually by speeding them up. Our definition that life is based on carbon can be extended: all life also possesses enzymes.

Further, all living things have an essentially similar structure: they exist as cells, a term introduced by Robert Hooke in 1665 when he noted that the cork he was studying was made of 'little boxes'. A cell is considered the basic unit of life: in the same way that a building can be taken apart until a pile of bricks remains, organisms consist of cells forming 'living bricks'. Each has, at minimum, a membrane to enclose a selection of chemicals.

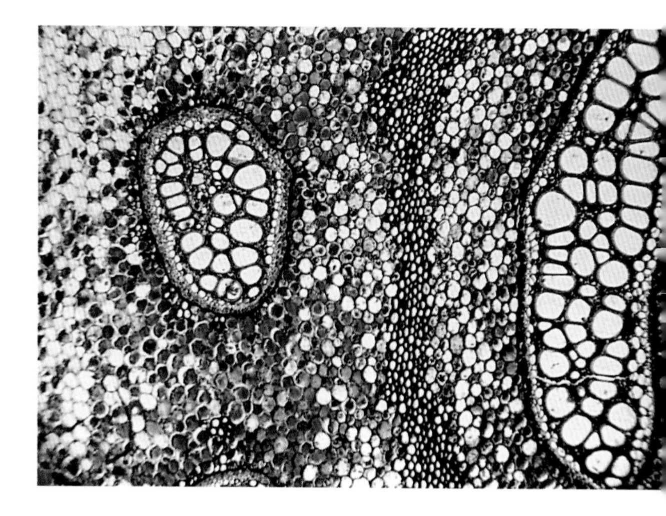

Cells are the building blocks of all forms of life. These cells are part of a rhizome, an underground stem which is used to store food for some plants. The large structures transport dissolved sugar and water. Cells were first identified in the form of 'little boxes' by Robert Hooke in 1665. Although most cells are minute, some algal cells may grow to be several centimetres long.

In effect, then, an alternative definition of life is possible: at its most basic, anything which is cellular in nature and possesses nucleic acids and proteins similar to those found in other organisms must be alive. The action of enzymes forms the clincher. It is therefore possible to define life without resorting to subjective assessments of what an organism can be observed to do.

Where do viruses fit into this scheme? On the border between being considered alive or only existing as complex chemicals, different doctrines argue with each other over their status. A virus is based on a protein and nucleic acid, can mutate and evolve, but there is still room for doubt. This is but one example of the problem in defining life: difficulties await the good doctor McCoy each time he lands on a new planet.

Beyond this discussion, it seems amazing that what we recognize as life should have so fundamental an attribute in common: an identical chemical basis. Yet, if the theory of evolution is accurate, it is a reasonable observation. Old beliefs of spontaneous generation,

wherein maggots arose within decaying meat rather than being involved with the cause of decay, were finally disproved by Louis Pasteur in the 1860s. If natural selection in a cause and effect chain, rather than spontaneous generation, is the key to change in species, the logical conclusion is that all life must have arisen from a single set of chemicals – which must themselves have arisen by spontaneous generation. A mechanism for this concept of the origin of life is therefore required.

In the 1920s two separate theories of how life began were presented by Aleksandr Oparin and John B.S. Haldane. Oparin, a Russian, suggested that high temperatures would have been sufficient to form molecules from elements, and in turn more complex chemical compounds from those molecules. Haldane became interested in forming mathematical models of evolutionary change and mutation rates, and proposed that ultra-violet radiation from the sun was sufficient to power the process of forming new chemicals, in particular from water, carbon monoxide, ammonia and methane. These initial reactions may have first taken place in earth's turbulent, primordial atmosphere, washed into oceans with a cocktail of ammoniacal rain and the sizzle of lightning discharge. Alternatively, perhaps the heat of a lava flow was required to bubble a thermal pool, where raw sulphur was also present.

Both theories – heat and radiation – depend on a lack of oxygen in the air. Oxygen is so highly reactive that it would block the formation of other compounds by reacting with their constituent parts. However, in this violent era oxygen was already locked to other elements, leaving the atmosphere filled with what we would consider poisons. A lack of oxygen also prevented the formation of an ozone layer (ozone is constructed of three linked oxygen atoms), enabling high-intensity ultra-violet light to penetrate the atmosphere. The proto-life arena was one of hot, steaming seas, broiling with volcanic ash and surging currents, becoming turgid with a vast array of chemicals which could, in turn, react with one another. No such process could take place today, not only due to the presence of oxygen but because any life-giving chemicals would be broken down by bacteria, using them as foodstuffs.

After experiments made by Stanley Miller in 1953, Haldane's ideas of life-giving chemicals forming under the power of the sun's radiation are thought more probable than Oparin's. Miller discharged sparks of electricity (simulating lightning, or high-intensity ultra-violet radiation) through a gas which simulated earth's atmosphere in the distant past, and formed carbon-based molecules in the water at the bottom of his flask. They turned out to be amino acids, the basis of proteins.

It is all very well to conjure images of a soup-like ocean, rich in new and exotic chemicals, but a means of growth and development is also required. In 1959 Norman Pirie theorized that this diverse set of chemicals would have produced more variety over a period of time, with simple structures randomly colliding to produce elaborate ones, a form of chemical evolution where compounds of increasing complexity might be formed. Eventually, a molecule might be produced which was capable of reproduction, an essential for living systems. The haphazard nature of chemical evolution would therefore be replaced by an ascending array of randomly forming biochemicals, Pirie thought; if he was correct, biological evolution was under way perhaps as much as 4,000 million years ago.

In order to grow, any life-form must obtain raw materials from its surroundings to manufacture new products, in particular more amino acids to make proteins. Only then can the organism reproduce: this would apply to a

Opposite: Life can thrive in an unbelievable range of conditions. The hot springs at Lake Bogoria in Kenya eject roaring gouts of boiling water with regular monotony, as do other volcanic vents across the world. The land surface is fragile here, and easily gives way; it is possible to boil eggs in the streams, which are littered with the bones of flamingos that have alighted in the wrong place. Yet, the super-heated pools of water hold living green plant filaments in perhaps the same manner as did earth's early proto-life arena.

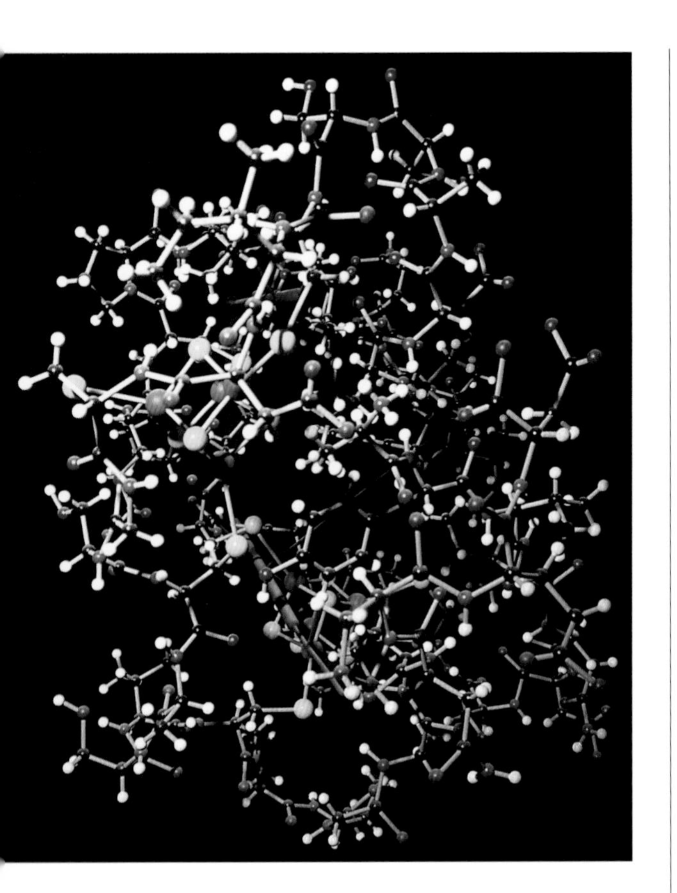

Even 'simple' forms of life require complex chemicals to grow. This is a representation of a single protein molecule called ferredoxin, which is constructed from amino acid units with iron and sulphur. It is involved in energy production in cells, including the most lowly of bacteria. The latticework of atoms is held together by carbon (the black components in the model), the element which all life relies upon as it binds together other elements such as oxygen, hydrogen and sulphur in an unbelievable variety of ways.

simple biochemical just as much as it would to a bacterium or bird. In addition, the organism must contain instructions which determine *how* the amino acids should be formed, and how a copy of itself should be made, something like having a construction diagram or blueprint to follow. In today's terminology, the essential biochemical is called DNA and is shaped like a long thread. Attached to this thread are genes, each of which incorporates specific pieces of information. Each gene might determine what raw materials are required to be joined together to make a section of protein, for example. At its most basic, the first 'living' biochemicals must have been able to perform like DNA.

There is a problem with the concept of such 'living' chemicals growing in the hot oceans of this early earth. As raw materials drift by they might be captured by chemical reactions and built into larger proteins following the instructions in these proto-genes. However, without a structure to prevent it, the newly made proteins would drift away. To enable any advance in evolution, the DNA must become enclosed in a container so that what it produced was retained.

In 1995 reports appeared of a Zurich-based research success. The Swiss team took small bubbles of fat called liposomes, into which they inserted very simple genes. The liposomes were then placed in water which contained a range of nutrients: the genes made copies of themselves. The presence of some chemicals caused new liposomes to appear spontaneously, or to reproduce.

What is especially interesting, and the reason for the Swiss team's use of liposomes, is that many ancient meteorites contain a percentage of carbon, some of which is organic and consists of complex arrays including paraffins, alcohol, sugars, amino acids – and liposomes. Earth currently intercepts 3,000 tonnes of 'cosmic dust' each year (estimated to have been 60,000 tonnes in the distant past), of which 300 tonnes is organic. Sidestepping the question of how these organic molecules could have formed and been locked within a meteorite, could this be an alternative, or additional, source of the chemistry of life? Perhaps here lies the answer to the structure of the first cell; perhaps life on earth began, as Sir William Thompson thought, with 'countless seed-bearing meteoric stones'.

Whether liposomes formed and captured the proto-genes, or vice versa, is unknown. Certainly, whatever form life first took, it would have been very simple, probably existing as short strands of gene-bearing DNA contained in simple, liposome-based cells. As these multiplied in number, they were protected from radiation by the oceans and were fed with raw materials washed from the air.

Oxygen is an extremely reactive gas, so it easily combines with other elements to form compounds. If compounds are broken down to simpler forms, energy is released. What was left over after the energy was extracted by the cells was waste, which in this case included oxygen. As the reactive elements of earth were slowly used up and were trapped in rocks, oxygen was finally released into the air. It was therefore life itself which slowly changed the atmosphere, and around 600 million years ago the level of oxygen in the air reached about 1 per cent (today, in comparison, the earth's atmosphere contains about 21 per cent oxygen).

To early cells, this represented a vital variation in more ways than one. A changing atmosphere represented an alteration in the environment which spurred on the forces of natural selection: cells had to adapt to the new, 'poisonous' oxygen gas. On the other hand, the presence of oxygen permitted a leap in evolution which eventually led to the formation of four types of cell.

Oxygen can be efficiently used in a cell to break down large, carbon-containing compounds in order to extract energy. It is a controlled reaction, in some ways similar to burning a log: energy is released as heat and light while the wood is consumed in the presence of oxygen. With respect to cells, this reaction is called respiration. It is easy to surmise that, with oxygen available, some cells began to make use of it and evolve new and different abilities, while others remained unchanged and did not. This concept helps explain the basic difference between two types of organism (in particular, microbes) found today: aerobic (oxygen requiring) and anaerobic (oxygen not required).

Complex cells rely upon internal structures to accomplish different functions, such as providing a site for the manufacture of protein or the release of energy when large 'food' molecules are broken down into smaller ones; it is a similar concept to a factory divided into departments, each with its own machinery. These internal, functional structures are termed organelles. Modern cells contain organelles called mitochondria, which are the sites where energy is released during respiration. Mitochondria have a number of similarities to aerobic bacteria, not the least of which is the presence of (now unused) DNA, indicating the possibility of a prior, independent existence. At some point in the evolution of life, it appears likely that a large cell engulfed one of the early, aerobic 'bacteria', and began to use its capability of releasing energy by using oxygen. In time, the host cell and the bacterial cell became so closely associated that the combination became permanent: an animal cell had formed.

Chemical reactions are not the only source of energy: there is also sunlight energy. The process of absorbing sunlight and using it to transform raw materials (water and carbon dioxide) into sugar, a carbon-based, high energy content substance, is termed photosynthesis: 'synthesis with light'. This is a characteristic of plant cells, a prerequisite being that sunlight can first be absorbed. To do so, plant cells use an organelle called a chloroplast (named for its possession of chlorophyll, the chemical which absorbs sunlight). Chloroplasts, as well as mitochondria, possess the remnants of DNA and were probably once able to live independently. If one of the early animal cells engulfed a proto-chloroplast, able to use sunlight energy, the association would have produced an early algal cell.

Today's life-forms are all based on these cell types, even though there is a huge variation. Aerobic cells are the most common, although some anaerobic bacteria abhor oxygen to such an extent that they die if the slightest trace of it exists. They can survive in deep-ocean volcanic vents, while their evolutionary cousins have the freedom of the world.

Early life, promoted into new forms by the presence of oxygen and driven by the forces of natural selection, reproduced in a simple

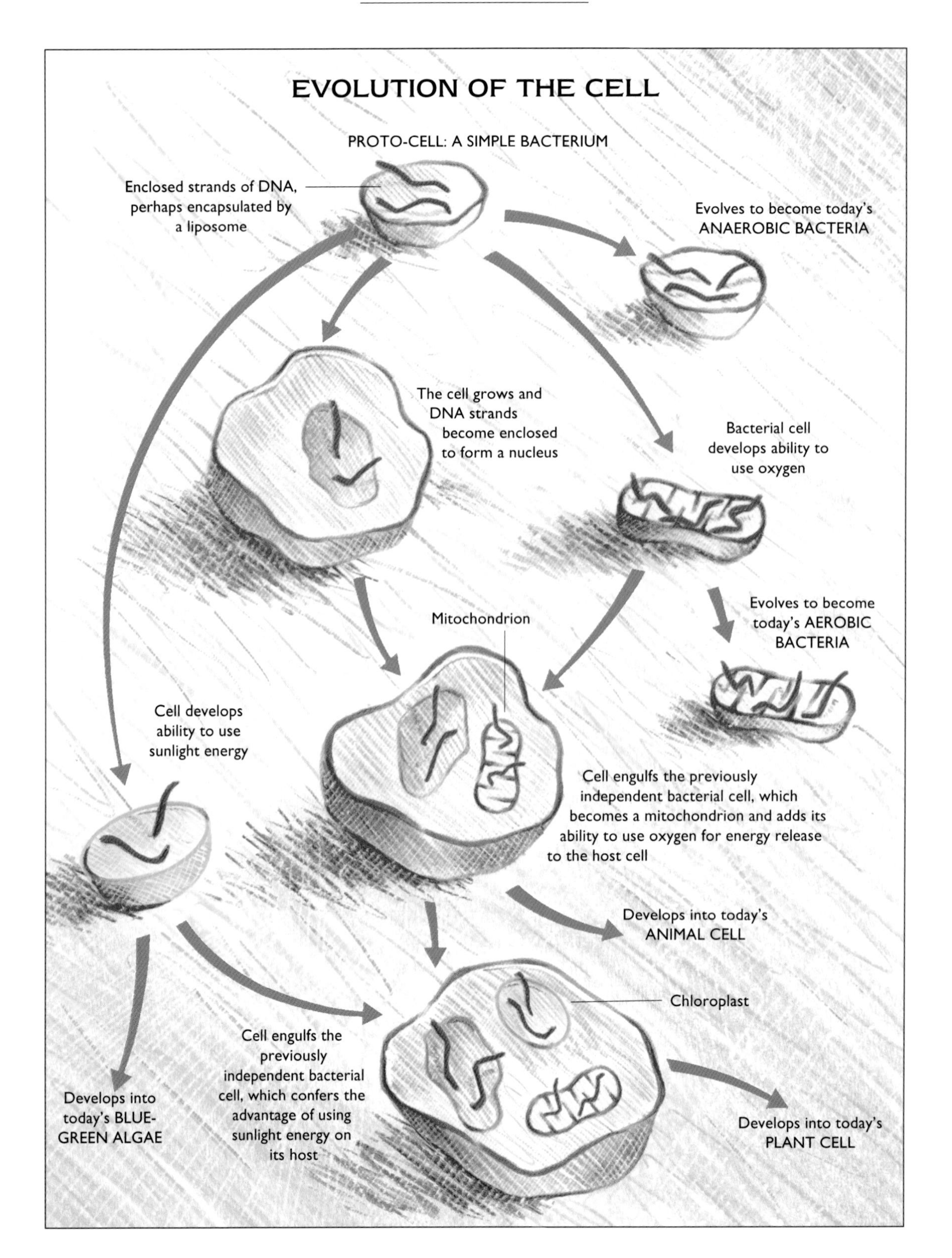
EVOLUTION OF THE CELL
PROTO-CELL: A SIMPLE BACTERIUM
Enclosed strands of DNA, perhaps encapsulated by a liposome
Evolves to become today's ANAEROBIC BACTERIA
The cell grows and DNA strands become enclosed to form a nucleus
Bacterial cell develops ability to use oxygen
Evolves to become today's AEROBIC BACTERIA
Mitochondrion
Cell develops ability to use sunlight energy
Cell engulfs the previously independent bacterial cell, which becomes a mitochondrion and adds its ability to use oxygen for energy release to the host cell
Develops into today's ANIMAL CELL
Chloroplast
Cell engulfs the previously independent bacterial cell, which confers the advantage of using sunlight energy on its host
Develops into today's BLUE-GREEN ALGAE
Develops into today's PLANT CELL

Bees can exhibit parthenogenetic characteristics: males develop from unfertilized eggs as drones, while fertilized eggs produce female workers. Here, workers are tending larvae contained in cells within the comb.

manner. To reproduce, a cell must first manufacture a new set of DNA (the process of replication) before dividing into two smaller 'daughter' cells, each with coded instructions identical to their parent's. Even when cells combined to form larger, multicellular structures, the ability to reproduce by simple means was retained: small pond animals called hydra grow miniature versions of themselves as buds, which are released when ready and, in any case, cell division is needed for growth. Corals can grow in the same manner. In another technique, a single cell is grown to maturity as a genetic copy of its parent; the process is called parthenogenesis (sometimes referred to as virgin birth) and is used by many insects, including greenfly and bees (for the production of male drones). In each case, the DNA has to be replicated.

DNA is a complex chemical. Making a copy involves a series of steps – any one of which can go wrong. Faults in replication can lead to disaster: the DNA no longer functions as it should, and the cell dies. However, faulty copying can also introduce random variations into the genetic code, and these 'mistakes' produce new characteristics in the affected cells. Any such variations were subject to Darwin's forces of natural selection, and produced the ability for organisms to better mould themselves to the environment. The process was slow. However, around 1,000 million years ago a vital step was taken: the introduction of sex.

Sexual reproduction, as a concept, is a strange one. On the one hand there are the perfectly operational copying techniques of cell division, budding and virgin birth. The result is a (theoretically) perfect genetic copy: a clone. The alternative, sexual reproduction, permits the exchange of DNA between two parents, which introduces new variations as genes are combined in fresh mixtures. However, half the individuals do not produce offspring; males, in production terms, are a total waste of space. In addition, half the so-called valuable genes available in the gene pool are wasted, discarded when the specialized sex cells (sperm, pollen or

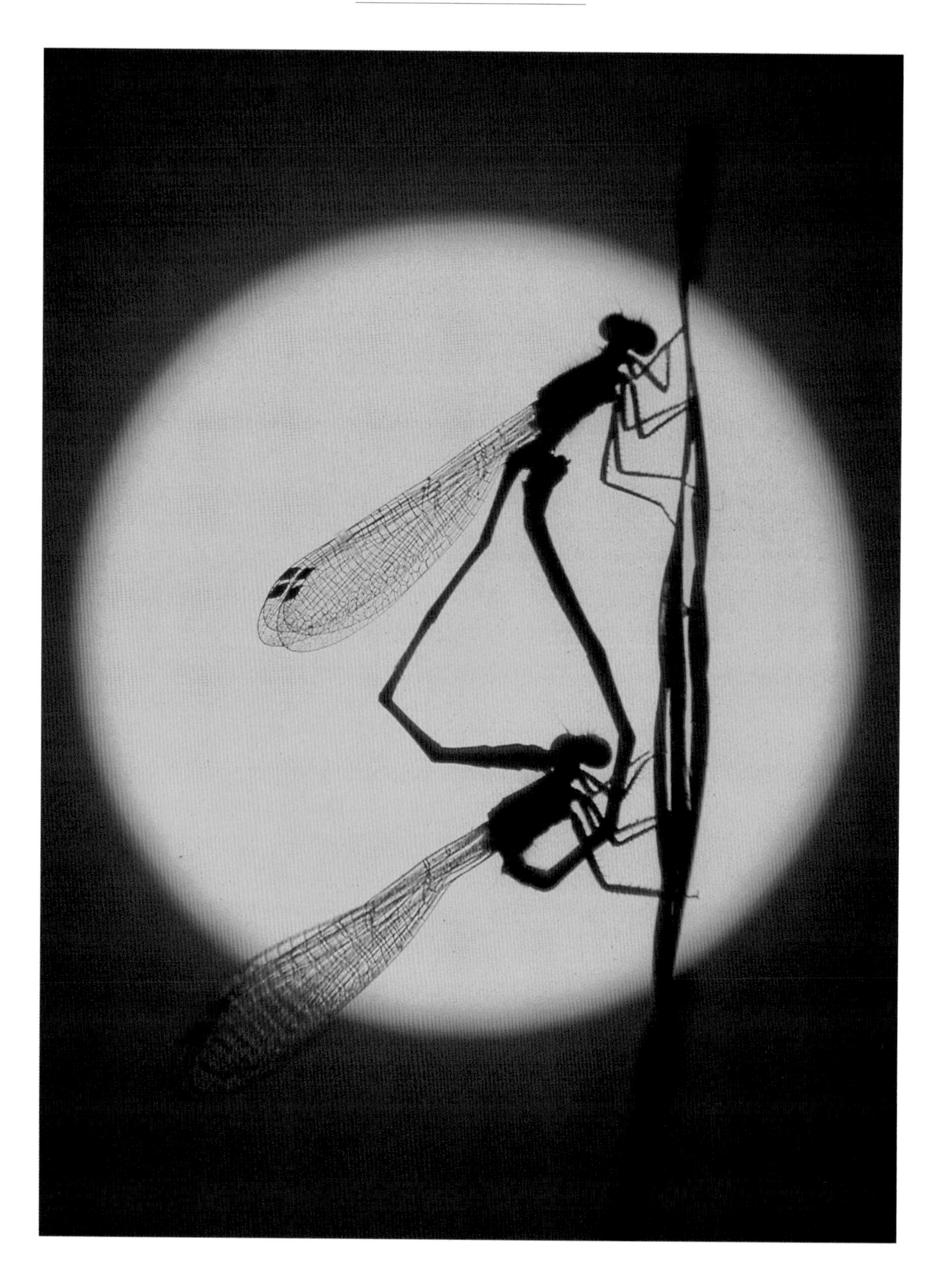

ova) are made. There must be a good reason why all the effort to develop variation leads to a system which disposes of half of its genes. In short, why does sex exist?

The answer may lie with several factors. Reproduction without sex (asexual reproduction) will pass on 'poor' genes (genes which may arise by chance, but which do not confer benefits to the organism) to the offspring as all the genes must, by default, be copied. To help avoid this, many asexually reproducing organisms make use of both sexual and asexual techniques, thereby having the best of both worlds. Bacteria, for example, are able to exchange strands of DNA and thereby reshuffle their genes, as well as reproduce by splitting into two new cells. The reshuffling of genes involved in sexual reproduction means that abnormal, damaged or 'poor' genes do not become concentrated; as they arise, they are diluted by new mixtures and eventually lost to the gene pool during the course of natural selection.

In 1995 the result of an experiment into the advantages of sex was published. A research team, led by Robert Dunbrack, used a measured quantity of flour to feed two populations of flour beetles in jars. The flour was laced with a low concentration of insecticide, enough to cause an adverse alteration in the beetles' environment, but not enough to kill them. Offspring borne by the two populations were treated differently: Dunbrack removed the young from one population, and replaced them with an equal number of beetles taken from a population fed with normal flour, while the second population was left alone. At intervals, the remaining flour was shared between the jars in proportion to the surviving beetles.

Effectively, the populations were competing for their food; one population remained genetically stable and simulated a virgin birth situation (the jar where the beetles were replaced) while the other was able to evolve. At first, the evolving group lost ground as they became affected by the insecticide, but after only five generations the resistant young which evolved were always able to outstrip the non-evolving population, which eventually died out. When it comes to evolution, sex permits a faster response to change.

Bacteria and other 'lower' forms of life, using asexual reproduction, can reproduce extremely quickly. As with all parasites, the life cycle of disease organisms is shorter than that of the species which they attack, and this factor alone means that new variations can spread with speed. While bacteria can reproduce every twenty minutes, even those insects with short life cycles require weeks or months – and mammals take years. It is estimated that 200 generations are enough to produce a new species: bacteria can accomplish this in 66 hours. Even without forming a new species, this rate would quickly, by blind chance, allow bacteria to evolve into a new strain against which a slower, asexually reproducing host would have no defence. The same argument applies to parasitic organisms: an asexual host cannot respond quickly enough.

The advantage of mixing DNA from two sources during sexual reproduction, however, arguably confers a more advantageous production rate of new variations. As variation is fuel to the process of natural selection, the speed of evolution is therefore enhanced. There is an underlying reason for using only half the available genes at each reproductive step: far from wasting them, unused genes remain in the organism to be presented to the next generation. What is of no use to a species today may prove tomorrow's salvation. Sex not only shuffles genes as a means of staying ahead of disease, it also causes unused genes to be stored until such a time when they might become more useful.

Speedy sex (or, to be more accurate, an enhanced evolutionary rate due to sexual reproduction) may be what has contributed most to our biodiverse world. For whatever reason, and in more ways than one, there seems to be a link between sex and the spice of life.

Opposite: Sexual reproduction offers an opportunity for shuffling genes between individuals. The offspring of these mating damselflies will possess a mixture of genes which differs from their parents and encourages the production of new variations.

C H A P T E R   T H R E E

THROUGH THE LOOKING GLASS

' 'Twas brillig, and the slithy toves did gyre and gimble in the wabe . . .'
'That's enough to begin with,' Humpty Dumpty interrupted: 'there are plenty of hard words there . . .'
'I see it now,' Alice remarked thoughtfully: 'and what are "toves"?'
'Well, "toves" are something like badgers – they're something like lizards – and they're something like corkscrews.'
'They must be very curious-looking creatures.'
'They are that,' said Humpty Dumpty.

Lewis Carroll, *Through the Looking Glass*, 1872

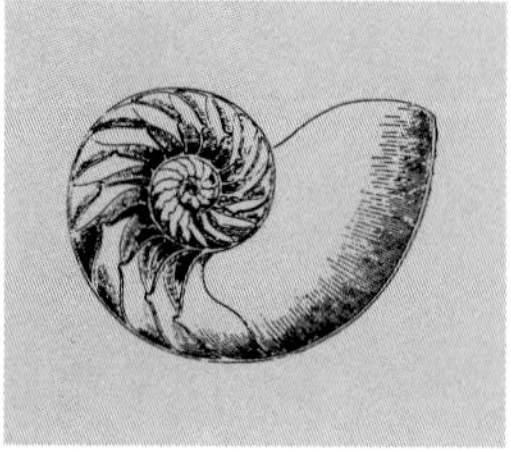

LINNAEUS MADE AN outstanding advance in biological studies when he introduced his binomial system but, even though it is still in use today, there have been inevitable difficulties as we learn more about our world of fellow plants and animals. In particular, even the refined categories which permitted a subdivision of Linnaeus's 'worms' could not withstand the pressures of a century of collecting and knowledge that species are not, after all, immutable. As the introduction to the 1880, six-volume *Cassell's Natural History* noted:

> *It must be remembered . . . that the best classification is but an attempt of a finite understanding to arrange the infinitely variable things of Nature. It is but an artificial and arbitrary arrangement which is necessary for study.*

Linnaeus and successive biologists used a basic classification of two groups: animals and plants. However, some forms of life proved so dissimilar to Linnaeus's 'accepted' kingdoms that alternatives were needed. In fact, several systems of classification exist which place some of the more 'difficult' life-forms (including those of bacteria, fungi and single-celled organisms) in different relationships to each other, probably the commonest being the five-kingdom system.

In the 1950s R.H. Whittaker divided living organisms into groups based on their cellular arrangement, number of cells, and the method of nutrition. Cells formed two types: prokaryotes (single, simple cells, such as bacteria and the blue-green algae, which do not contain a nucleus and enclose DNA in the form of a loop) and eukaryotes (cells with more complex internal arrangements, containing their DNA inside a nucleus). Methods of nutrition were separated into three types, depending on whether food was absorbed directly from the surroundings over the surface area of the organism, was ingested as ready-formed chemicals, or was manufactured using sunlight energy, carbon dioxide and water (the process of photosynthesis).

Fungi, for example, were previously classified in the kingdom of plants; after all, they had the same range of movement, reproduction and growth as other plants. However, their mode of nutrition differs markedly; they are not green and do not photosynthesize. Some single-celled organisms possess properties of both animals and plants. For example, *Euglena*, a genus of organisms which can cause a red or green colour to appear in pond water, are green and can photosynthesize and were therefore once classified as plants. Equally, they have a gullet for ingesting food and can swim and move in animal-like ways.

Whittaker therefore proposed five kingdoms: Plantae and Animalia, plus the Monera (to include bacteria and blue-green algae),

Fungi derive their nourishment from organic material, such as rotting wood and leaves, unlike plants which obtain their energy from sunlight. The part of a fungus most often seen, as a toadstool or mushroom, is its fruiting body, used in reproduction. Here, *Lycoperdon perlatum* puffballs are ejecting spores, each of which is capable of infecting a new area.

Coypu (above; also known as Nutria in the USA) and marmots (left) are sufficiently alike to belong to the same order of mammals, the Rodentia, but nevertheless are different enough to be assigned to the Capromyidae and Sciuridae families respectively. All rodents are believed to have arisen from a beaver-like ancestor which evolved in the Palaeocene, around 60 million years ago.

Protista (some single-celled algae and protozoans, previously grouped as single-celled animals) and Fungi. Within each kingdom, as before, groups of greater and greater similarities may be found. The group representing the basic body plan is named the phylum (there are about 35 of these), which is subdivided into a class, order, family, and finally genus and species.

The essence of classification remained unchanged. To classify any new organism, Linnaeus's established system required that characteristics were observed and counted:

Discovered in 1859, *Welwitschia mirabilis* presents some of the problems involved with using physical features to assign an organism to its 'correct' classification group. Just as the platypus exhibits a contradictory range of features, Welwitschia possesses elements normally associated with club mosses, flowering plants and coniferous trees.

Living exclusively in the arid Namib Desert of south-western Africa, Welwitschia consists of only two leaves (they appear to be more numerous as they are split into sections) and is dioecious (plants are sexed: each is either male or female, the latter bearing cones as reproductive organs). Individual plants have been dated to over 1,000 years old.

how many stamens does a flower possess, does an animal possess a backbone, how many legs or wings are there? As an example, the animal kingdom contains a phylum of similar organisms bearing a backbone: the Chordata. Within this is a class of warm-blooded, hairy animals: the Mammalia. In turn, the mammals are divided into orders with yet more characteristics in common, producing groups such as the Carnivora (which includes the cats, dogs and bears) and Rodentia. A subdivision of Rodentia produces families of animals, for example the Castoridae (beavers), Sciuridae (squirrels, marmots), Capromyidae (coypu) and Muridae (rats, mice, lemmings and voles). Eventually a group with a great deal in common is attained with the genus, of which individual members are termed species, for example, *Marmota flaviventris*, the Yellow-bellied Marmot.

This is a type of classification system termed 'phenetic'; it depends on similarities to place organisms into groups. These might be based on external features, internal anatomy, or even the presence or absence of biochemical compounds. As it is so useful in tabulating organisms, and so easy to use in identification, the binomial system is the commonest form of classification, but it is not the only one in existence.

A second system is evolutionary classification, which may use some of the data revealed by a phenetic system but depends for the most part on how organisms relate to each other in evolutionary terms. For example, how did birds arise as a group? Using evolutionary classification, we know that birds have some of the characteristics of reptiles (their legs are scaled, for example) and therefore we believe that they arose from reptilian ancestors. Evolutionary classification helps plot the ancestry of animals and plants, placing the original, simple bacteria and algae at the base of a spreading tree. As groups diversify, branches leave the main trunk and, in turn, give rise to twigs which subdivide again and again to end in modern-day species: in analogy, the species appear as leaves. Even so, the analogy is poor and the system is prone to producing fallacies, but it is a useful technique in plotting fundamental relationships using evolutionary evidence.

An example would be to consider birds and mammals: did they arise independently of each other or did, perhaps, birds evolve from an already developing mammalian line? Where, exactly do the branches leave the tree, or are they aspects of the same limb? Here, anatomy holds a clue.

If a study of mammalian bone structure is made, concentrating on an arm and hand, identical numbers of bones are found in the same relationship with each other in every case, be it a bat where 'fingers' support the wing, or a horse where a hoof (the toenail) appears at the end of the limb. A reasonable deduction is that all mammals share a common ancestor, after which evolution has permitted a divergence of form so that different bones attain different shapes and functions. This is known as divergent evolution; organisms are growing apart from each other in form and function. Convergent evolution occurs when organisms attain a similar form and function, though they have arisen along different routes: birds and bats are both able to fly, although they followed separate evolutionary pathways. Did a whale, a mammal, evolve in the sea while its evolving descendants migrated to land, or did it originate on land and migrate to the sea? A study of whale flippers shows that it has the same number of bones as a land-bound animal, indicating that it took the latter route.

Evolutionary classification is not a perfect science. If a comparative study is made of the heart of an amphibian, reptile, bird and mammal, differences in the chambers, and in the major blood vessels which attach to the heart, are readily apparent. These changes enable a sequence of events to be constructed which imply that birds and mammals have a common ancestor, and that birds attained their characteristics via a route which extended via amphibians and then reptiles, but that mammals arose directly from amphibians. However, we also have fossil evidence ('hard' evidence in more ways than one) which tells us positively that mammals arose directly from reptiles. Inferences made through a study of anatomy, and other deductions, have therefore to be taken with care. Even so, it has proved possible to construct an evolutionary classification which indicates the relationship of the major groups to each other, making a branching ladder of complexity. It is here that the true scope of biodiversity – both in form and numbers – is revealed.

Of the five taxonomic groups, the number of Monera (the bacteria and blue-green algae) which have been identified total something approaching 5,000 species, the Protista (the single-celled organisms) manage well over 30,000 species, while the array of Fungi – including all the slime moulds – attain a staggering 70,000 species. Even so, the Animalia and Plantae offer even greater numbers.

The most basic division in the modern-world Animalia is into vertebrates and invertebrates, the latter forming an incredibly diverse series of groups. Even here, we have much to learn: as recently as 1995 the discovery of a new body form, the Cycliophora phylum , was claimed on the basis of a single species, *Symbion pandora*, found growing on the mouth of a lobster – the announcement being the zoological equivalent of man first walking on the moon or learning to fly.

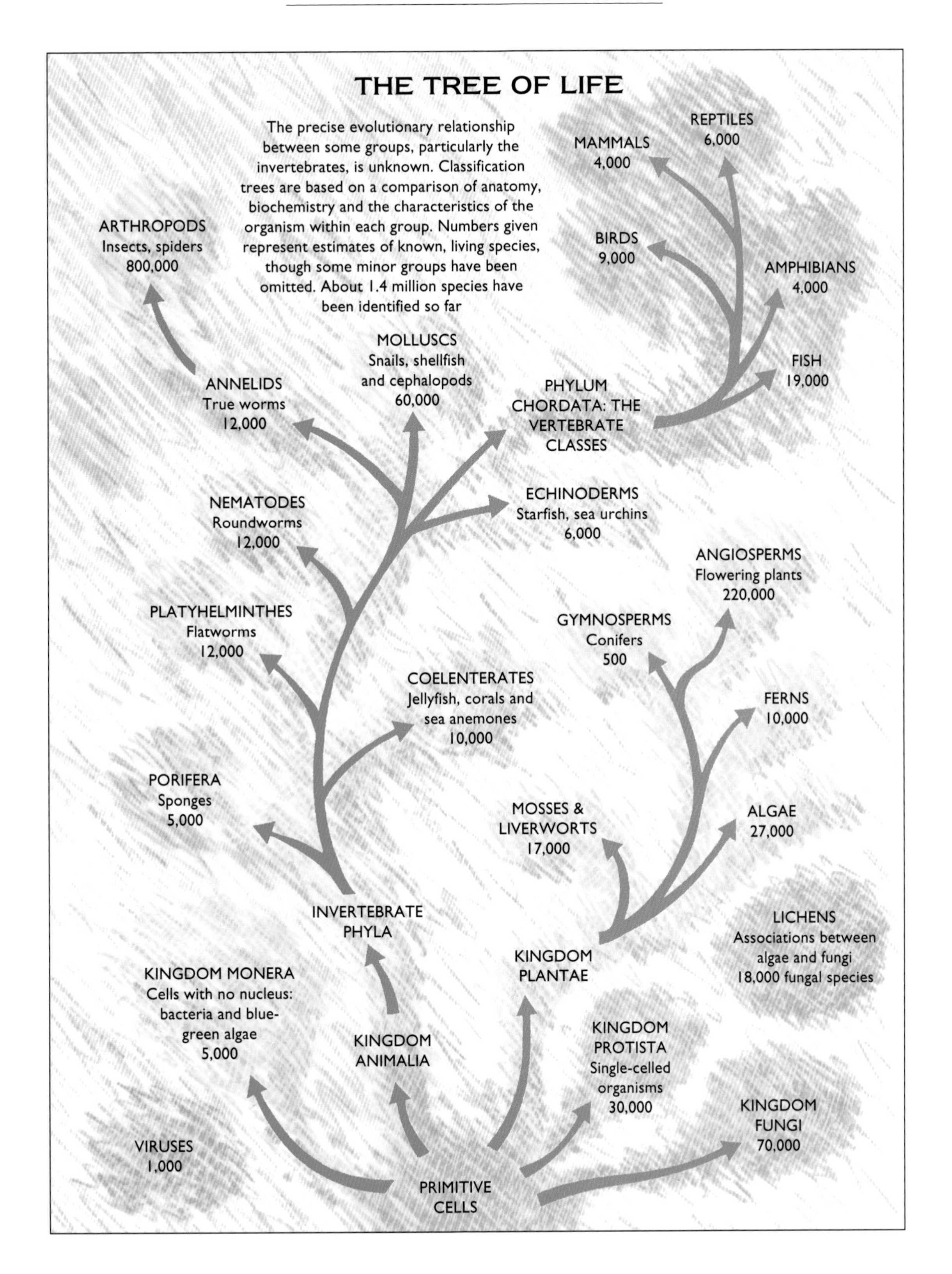
THE TREE OF LIFE
The precise evolutionary relationship between some groups, particularly the invertebrates, is unknown. Classification trees are based on a comparison of anatomy, biochemistry and the characteristics of the organism within each group. Numbers given represent estimates of known, living species, though some minor groups have been omitted. About 1.4 million species have been identified so far
ARTHROPODS
Insects, spiders
800,000
MAMMALS
4,000
REPTILES
6,000
BIRDS
9,000
AMPHIBIANS
4,000
FISH
19,000
MOLLUSCS
Snails, shellfish and cephalopods
60,000
ANNELIDS
True worms
12,000
PHYLUM CHORDATA: THE VERTEBRATE CLASSES
NEMATODES
Roundworms
12,000
ECHINODERMS
Starfish, sea urchins
6,000
ANGIOSPERMS
Flowering plants
220,000
PLATYHELMINTHES
Flatworms
12,000
GYMNOSPERMS
Conifers
500
COELENTERATES
Jellyfish, corals and sea anemones
10,000
FERNS
10,000
PORIFERA
Sponges
5,000
MOSSES & LIVERWORTS
17,000
ALGAE
27,000
INVERTEBRATE PHYLA
LICHENS
Associations between algae and fungi
18,000 fungal species
KINGDOM MONERA
Cells with no nucleus: bacteria and blue-green algae
5,000
KINGDOM PLANTAE
KINGDOM ANIMALIA
KINGDOM PROTISTA
Single-celled organisms
30,000
KINGDOM FUNGI
70,000
VIRUSES
1,000
PRIMITIVE CELLS

Of the more established groups, the Porifera phylum constitutes the base of the animal tree with around 5,000 known species, though at least one estimate places the real number at twice that. In some areas of warm, shallow seas the Porifera form some 80 per cent of all living matter and are rather better known as sponges. Living in water, sponges are an ancient group of great diversity, which probably arose in the oceans around 1,000 million years ago. Some form flattened areas of bright colour on the undersides of rocks, while others rear from the ocean floor in chunky, soft structures which may, in turn, provide hotels for other organisms to inhabit.

That sponges are basic organisms can be realized if one is broken up into its constituent cells: remix them, and a sponge begins to form from the wreckage. Single cells can therefore exist in the guise of a single organism with a fair degree of complexity, while also maintaining some sort of independence. Sponges possess feeding channels which funnel water through the organism, driven by ocean currents and the action of whip-like flagella. Support is attained using calcium or silica, the basic element used in making glass; if you use a real sponge in your bath it is this dead skeleton which soaks up water.

However, although sponges are multicellular, they do not have any form of nervous system. That honour belongs to the next phylum, which comprises the jellyfish, corals and sea anemones. These Coelenterata are also multicellular, but the cells will not reconstitute into a new organism if they are separated one from another. They have a mouth to take in food to a central cavity, and a nerve net. This is literally like a net or web, spread through the animal and capable of linking all its parts. Muscles operate tentacles and specialized cells can ensnare or sting prey. As a name, 'jellyfish' is an apt one as the animals are 94 per cent liquid: a jellyfish drifts or moves in jerky pulses through the ocean, but out of water it is a sorry spectacle indeed.

Opposite: Jellyfish belong to the coelenterate group of animals, sharing their classification with corals and sea anemones. Comprising 94 per cent liquid, they carry stinging cells which capture prey as they swim or drift through the ocean, and possess a coelum – a central, fluid-filled cavity – which provides the basis of their phylum name.

The corals, just one segment of the 10,000 or so known coelenterate species, are very similar to sea anemones or 'upside down' jellyfish. Many people see corals in aquariums and immediately think of flowers: animals they remain, but with their radiating tentacles the resemblance to an open bloom is often striking. In the same way that flowers can be classified according to the number of petals, anemones and hard corals (those bearing a hard secretion around them; some corals lack this protective home and are termed soft corals) are based on the number six: tentacles are in multiples of six, and so on.

Some corals are solitary, but others – the corals of fact and fable in their ability to form huge, tropical, ocean-calming reefs – are colonial; that is, the individual polyps live together to form huge masses of calcareous cups, often brightly coloured and covered with a delicate membrane. Little wonder that the fossil record of corals is good: the skeletal remains are relatively easily preserved. Think of Australia's Great Barrier Reef at over 2,500 km long, successive corals growing on their dead ancestors as oceans rose or the land fell, or the wide expanses of limestone rock which are formed from a variety of corals and shells. Indeed, the 6,000 living species of these Anthozoa are matched by another 6,000 fossil forms.

It is no coincidence that corals (and some anemones) prefer to live in the upper, sunlit portion of the ocean: they require sunlight to live. How can this be, when they are animals and therefore obtain their food from catching prey? The answer lies with organisms called zooxanthellae: small algae held within the tissues of the coelenterate which photosynthesize and provide the animal with up to 90 per cent of its nutritional requirements.

As animals increase in complexity, more internal organs are required to sustain them. The next branch of the evolutionary tree belongs to the Platyhelminthes – the flatworms, over 12,000 species of them. With the flattened shape and a decreased reliance on water comes an overhead: if food can no longer be absorbed over the animal's external surface, an intestine is required. This is not, as might be thought, a one-way tube running through the body, but a branching, blind-ended system penetrating all parts of the body. In some flatworms, notably parasitic tapeworms, the intestine is not well developed; after all, a flatworm can afford to be somewhat wasteful with food when it hangs in a host's gut.

Nervous systems in different species of flatworm may be simple or complex. Some have tiny brains, while others possess separate nerves and sensitive patches which approximate to eyes. This enables them to detect light and orientate themselves to move away into darkness, an evolutionary trend towards the sensitive capabilities of the 'higher' animals. The use of such terms, 'higher' and 'lower', sometimes misunderstood: there is no element of stating 'better' or 'worse', but only an indication that evolutionary classification has placed them higher or lower on the evolutionary tree. Neither should there be any misuse of the concept of 'successful', often applied to the larger groups: any organism which has survived and evolved to find its place in this biodiverse world must, by default, be successful.

Flatworms, with their minute brain, have also developed the ability to learn in a limited way. Experiments show that a flatworm can learn to pass through a maze. Of even greater interest, especially to supporters of Lamarck's theories that characteristics gained during life can be passed on to offspring, is the revelation that cells from a destroyed flatworm, fed to a companion, will confer the ability to pass safely through the same maze.

With a flattened body, no longer able to absorb food through the organism's skin, came the requirement to transport food, via an intestine, to all parts of the body. As bodies become thicker, the problem intensifies: oxygen as well as food must be supplied to all cells. The 12,000 known roundworms, or Nematoda, begin a trend which continues into the true worms, or Annelida. This group includes the earthworm and its relatives, such as leeches.

Earthworms, which bear many ringed segments along their body (the term 'annelid' is derived from the Latin word for a ring), exhibit a number of more complex characteristics when compared with the 'lower' groups. In this progression up the evolutionary tree, there are now strong sets of muscles, excretory structures in each segment, a good circulatory system (there are individual blood vessels) with a simple pumping structure (the heart), and a mass of nerves making a clump near the worm's 'head', acting as a brain. The concept of 'head' is used very loosely; simply, it is the end which takes in food and contains the brain. It is also the area of oldest segments, because new segments are always added at the 'tail' during growth.

The segments are incredibly important to the worm's design as they form a series of rings with internal divisions, like a set of bulkheads in a submarine. In fact, the analogy is quite good if, instead of an air-filled submarine, you imagine a liquid-filled worm with small 'doors' connecting between sections. It is the liquid which gives the worm its support (a hydroskeleton) and provides something for the muscles to push against, allowing movement. The segments can also become specialized, for example in reproduction. The aquatic palolo worm carries sexual cells in its tail segment, which breaks off under a full moon only in November. The masses of segments float upwards from their Pacific reefs, and mingle at dawn on the surface: sexual reproduction at a distance!

Possessing segments helps explain the gardener's confidence that if an earthworm is cut in two each half lives on as a new worm, as at least the 'bulkheads' avoid a total loss of liquid. In fact, only the front half of the worm (which lives for up to 15 years) regenerates with any certainty. The palolo worm also regenerates its tail after losing its posterior, though other species of worm may not. If you are a gardener you will also appreciate the incredible numbers of earthworms which exist. There are some 12,000 species in the annelid group, but this pales into insignificance when the numbers of individuals are considered: in fertile grassland there may be up to 600 million worms in every square kilometre.

The development of the worm's internal structure was crucial in evolution. A rudimentary brain and nervous system (with some species possessing well-developed eyes) offer increased sensitivity, there is a proper intestine with digestion taking place outside the cell, to say nothing of excretory systems, a circulatory system, reproductive structures, and the segmentation which aids an internal organization. Blood carries oxygen to all parts of the body, though some species use chlorocruorin rather than iron-based haemoglobin in cells and, as a result, have green blood rather than red.

Hydroskeletons, as might be imagined, have limitations. In particular, there are restrictions on size: sea anemones can grow with the support of the surrounding water, but land-based organisms living in air lack the same degree of aid. In addition, while hydroskeletons permit movement, they do not offer protection against predation or water loss. To fulfil these functions, a new design is required: an external case. This container, termed an exoskeleton (an external skeleton) takes different forms, and is used as an important characteristic in classification.

The possession of a shell confers a rich opportunity for diversity: so far upwards of 60,000 species of Mollusca – oysters and clams, snails and slugs – have been identified (some estimates reach 100,000), with another 35,000 extinct species. If that outstrips the fossil records of other groups, it is because shells can be easily preserved; soft bodies do not fare so well as hard ones when it comes to being buried and preserved by minerals, and it is amazing that any soft-bodied jellyfish have been recorded in stone at all.

Molluscs are characterized by their possession of a shell made from a substance called conchiolin plus calcium carbonate (calcium carbonate is better known as chalk, itself formed from the skeletal remains of shelled sea creatures). Because these constituents and the method of their secretion can vary, a wide range of colour and texture is produced in shells. In the same way that trees can be aged by counting rings in the wood, some shells will reveal the age of their owner. Tree rings are dense and dark where growth is slow or has ceased, such as in the cold of winter when there are no food-producing leaves on the tree, and lighter and wider when growth is most rapid, in the spring. The same forces are at work in shells; cold weather can also interrupt growth and produce darker rings.

Shells are secreted by the animal's mantle, which also covers the gills and internal organs. The mantle in some species can be 'irritated' if anything gets inside it, such as a grain of sand, and 'shell material' (actually a crystalline form of calcium carbonate: mother-of-pearl) is used to coat it. Although pearls to make jewellery are usually obtained from oysters, they can be found in a range of species.

Shells are found in a wide array of shapes and sizes, from long, curved tubes to whorled cones. At first glance these cones and spires are similar, but differences are soon discovered: some, for example spiral in a right-handed direction (clockwise) while others, the minority, are left-handed.

Molluscs use the chemical haemocyanin in their blood cells, producing a blue colour. Movement relies on a muscular foot which ripples in waves of activity, and terrestrial slugs and snails leave behind a slimy layer of mucus as a trail. If it is the presence of a shell that, in part, defines a mollusc, where is it in slugs? Diversity again: the shell is reduced to a small, curved plate inside the animal (sea slugs have lost their shells altogether), in a similar fashion to the molluscan class of cephalopods: squid, cuttlefish and octopus. Here are, to some people, the monstrous animals of nightmare. Giant squids rear, kraken-like, to ensnare ships and pull groaning timbers below. The Luska still lives in Bahamian folklore as a giant creature residing in submarine cave systems, and boats do disappear – but only at the whim of incredible suction currents as the tide rushes through the caves. Then there is the octopus, with eight tentacles (squids have ten) and a reputation for intelligence and hunting out divers, dragging them to their doom.

So is there any truth in cephalopod stories and beliefs? An octopus can indeed learn to recognize objects and places, and can be trained to perform simple actions; the term 'cephalopod' links to this ability, from the Greek word *kephale*, meaning 'head'. Its brain has to be large and well developed, for an octopus's sense organs are extremely advanced and provide a wealth of information for it to deal with. The eye, for example, is capable of seeing more detail than our own. Giant squids have been found, and preserved specimens possess a body length of 7 m plus another 10 m of tentacles. Increasingly, new species are being found in the deep oceans and, at a depth approaching 2 km, a giant squid with a total length of over 24 m has been recorded in what is emerging as one of the world's most prolifically biodiverse habitats, alongside the rainforests and coral reefs. The giant octopus seemingly remains a myth, however. The body length of the Pacific Coast *Octopus punctatus*, one of the larger known species, rarely exceeds 35 cm, though its arms can catch prey up to 5 m away.

This specialization within the molluscs seems to be evolution heading into a corner: it appears that further advances are likely to be limited, and there are only about 400 modern species of cephalopod compared with over 10,000 fossil types. The cephalopods are the end of a long line of random development which began (fossils tell us) 1,000 million years ago, along the way passing and losing those most fascinating of organisms, the ammonites. Ammonites produced massive shells, so huge (some attained a diameter of several metres) that it seems they could have never moved. However, fossils can do more than give us the outside shape of an animal or plant: sometimes, the internal structure is preserved and we know that, like worms, ammonites had 'bulkheads' throughout their spiral shells, many compartments being air filled and therefore providing buoyancy. The ammonites specialized and died, or adapted and evolved onwards, 100 million years ago, leaving us with the remnants of a once-prolific, biodiverse ocean. A single, shelled genus remains with the Pearly Nautilus.

In a similar fashion, the next major group – the Echinodermata – now contains comparatively few species (only about 6,000 in all), though they are well known to every seaside child with a bucket and net: the echinoderms include the starfishes, sea urchins, sea cucumbers and all their relatives.

It is commonly believed that starfish possess five arms, but this does not apply to every member of the group. In the same way that the corals are based on the number six, the echinoderms are based on the number five – but this can occur in multiples. Brittle stars, with long, slender, highly mobile arms which give them a racing turn of speed, may have ten arms; the feather stars and sea lilies (grouped within the echinoderms as the crinoids) have even more. In contrast, it seems as though sea cucumbers have no arms, but their bodies bear five lines of tube feet; underneath the spines of sea urchins there is a rounded sphere which also carries five radiating lines of tube feet, the equivalent of the five arms of the common starfish.

Tube feet project through small pores in the animal's exoskeleton, each ending in a sucker similar in appearance and action to the structure which holds a mirror to a glass surface in a car. Powered by water pressure from within the animal, tube feet possess incredible strength – enough to tear apart the two shells of a mussel to reach the succulent food within, and to walk on. Here again there is proof of diversity in action: hydro-power for movement, with an exoskeleton casing. Every possible ramification of the 'five-based' design has been tried, tested, adapted and then tested once more.

The final major group of invertebrates (there are several other minor ones) is the huge collection of organisms making up the Arthropoda. Well over 800,000 species are classified here, around 80 per cent of all known animal species. Individual arthropods are highly specialized, and supremely successful in their adaptation to whatever ecological niche they inhabit. They all have jointed legs and an external skeleton, and can move on land as easily as they do underwater. Here are the crabs, spiders, ticks, mites, centipedes, millipedes, scorpions, horseshoe crabs and, the largest collection of all (750,000 species), the insects. Gone from our world is the immense group of the trilobites – all 4,000 species of them, lost to extinction at the end of the Palaeozoic era 250 million years ago.

Arthropods such as this Bahamian land crab possess a hard outer case which provides protection against desiccation, and individuals will only return to the sea to lay eggs at full moon when the tides have their greatest range. Although this waterproof exoskeleton enabled crabs to leave the sea, it also introduced a restriction on growth, which has to take place in spurts when the animal moults.

Perhaps the reason for the arthropods' colossal success is linked with this combination of a strong, outer casing operated with bands of muscles, plus a highly adapted internal organization. Body organs can be attached to the inside wall of the skeleton, leaving the centre to be filled with water, blood or muscle, or whatever the environment dictates. Appendages form legs and pincers; whatever is required, evolution has fashioned. Crabs colonize the ocean, some becoming land crabs which only return to reproduce when the tides are greatest, and eggs and sperm have the maximum chance for dispersal. Eight-legged spiders and six-legged insects proliferate over their terrestrial environment, adapted to extremes so that they need never approach water again although, through the course of evolution, insect species have returned to fresh water numerous, independent times.

Yet there is one problem with their design: the skeleton which contains and seals in body fluids, sometimes using an additional layer of wax, is a potential stranglehold on growth. How on earth does an organism increase in size when its outside skin is a hard, rigid structure?

Growth in crustaceans (crabs, lobsters and suchlike) takes place in spurts, new exoskeletons forming under the old. When the time is ripe the old carapace splits and is shed, the new, still-soft skeleton expanding and hardening to its armour-like consistency. Then the cycle begins again: the old moult casing may be eaten to reclaim some of its valuable salts, food reserves are built up and internal organs prepare for another change in life before another moult can be completed. Insects rely on passing through a series of larval stages which lead to the adult form, such as a caterpillar passing through a pupal stage to emerge as a butterfly.

Virtually all insects are characterized by wings; they have truly conquered the air – and did so some 350 million years ago. The aerial mastery gained by the insect group enabled them to take the planet by storm: dispersal was easy, food could soon be located in new areas, and escape from ground-based predation became possible. While insects are not indestructible, some are nearly so – at least in comparison with us humans. They fly to immense altitudes and survive the same degrees of air pressure and low temperature which are encountered by high-flying aircraft: some larvae can stand immersion in liquid nitrogen, while others can be boiled for short periods.

It is often quoted that a bee is not aerodynamic and, physics tells us, cannot fly. Yet fly it does, with the utmost efficiency. Any student of flight knows that wings require an aerodynamic surface; it must be curved to force air to take different paths under and over the wing, creating suction above by changing the air pressure as it flows. Bee wings – and those of many other insects – are flat at rest, but curve during flight along pre-ordained stress lines controlled by the strengthening veins.

Just having the right shape of wing is not enough: the wing must move not only up and down, but in a figure-of-eight motion. To retain efficiency, a hinge system is used whereby the whole exoskeleton is flexed by muscles to flip the wing through its cyclic routine. A huge amount of energy is required, hence the additional pumping action of the abdomen to help push blood around the body and obtain oxygen – the blood bathes the organs, while air is sucked from the outside through small tubes. Midges can attain 1,000 wing-beats a second using this system (a remarkable adaptation as this oscillating system is faster than nerve impulses), while hawk moths are the fastest fliers at 56 km per hour. Initial lift is gained without the need to run to create airflow over the wings (the reason why aeroplanes use a runway for take-off) using what is known as a tarsal reflex: houseflies, for example, do not fly unless they are already airborne. To take off, insects jump in the air and, as the foot-controlled inhibition which prevents flight is removed, the wings go into action. Altogether, insects have evolved an ingenious system of flight.

As members of the Chordata phylum, which includes all the vertebrates (fish, amphibians, reptiles, birds and mammals), birds have tackled the problems of flight in a different manner, using feathers and raw muscle power to operate very light wings. Their evolution is linked to the time of the dinosaurs, that extinct group of animals which has captivated the imagination of schoolchild and adult alike, spawning films such as *Jurassic Park* and a multitude of museum exhibits and rubber toys. The clue to their evolution came

Opposite: Sea squirts, such as this British Light-bulb Sea Squirt (*Clavelina lepadiformis*), are more correctly termed ascidians. Each animal, as a larva, swims through the ocean to a suitable location (this specimen was living beneath Swanage pier) and attaches itself head-first, then feeds by filtering water. They form interesting additions to the classification tables, as ascidians possess a hollow, dorsal nerve cord. They therefore have a link to all vertebrates: having a rudimentary backbone, they are classified within the Chordata phylum.

in the form of a fossil which was first discovered, somewhat opportunely as the 'missing link' aided acceptance of his theory, only two years after Darwin revealed his thoughts in *Origin of Species*. *Archaeopteryx* had reptilian features but also a feathered wing, which could have been used to glide and hence was a step towards powered flight; other fossil winged species have since been found in China.

Cold-blooded animals possess the same body temperature as their surroundings, while warm-blooded animals (birds and mammals) are maintained at a constant temperature (but therefore, on a hot day, a cold-blooded animal can attain a higher temperature than a warm-blooded one). Here is another example of diversification and a massive advance in evolution: once warm-blooded animals had developed, they were freed a little more from their environment and could evolve radically different internal behaviour. Research indicates that dinosaurs themselves may have possessed some means of regulating their internal temperature, a step towards true warm blood, and with this it becomes easier to see the route evolution took to attain the mammals and birds; birds may represent a modern-day dinosaur.

The issue of warm blood is crucial. All animals require energy to operate muscles for movement, and even to digest food. This energy is released during respiration, which is a form of chemical reaction. Cold-blooded

The Green Tree Python of New Guinea, being a cold-blooded reptile, must warm itself before it can become active. Its behaviour, as well as heat from the environment, dictates its body temperature.

animals, including all the invertebrates, fish, amphibians (such as the frog) and reptiles (for example crocodiles, lizards and snakes), cannot move with more than sluggish speed, or fulfil any bodily function, without sufficient temperature to permit the chemical reactions to proceed. Cold-blooded animals are therefore totally reliant upon the weather, unless they have developed special mechanisms or behaviour. Bees can raise their internal temperature well above their surroundings using shivering techniques, unlinking their flight muscles from their wings, to create friction. The birds and mammals, though, do not have to rely on behaviour such as basking in the sun, or other methods of sidestepping problems, as they have a constant supply of heat (the down side of the system is that an increased, and constant, supply of food is required).

Birds form the second largest vertebrate group with over 9,000 species so far identified, with fish taking the prize at more than 19,000 (though some estimates place them at over 30,000 species, including about 1,000 sharks). Even so, the numbers of vertebrates are far less than those of the invertebrates, a factor which increases when the amount of study devoted to various groups is considered: we probably know most of the mammal and bird species, but by far the majority of invertebrates remain to be discovered. A problem lies not with obtaining an organism which may be new to science, but in easily taking it through the series of checks against other species which is required before it can receive its Latin name: the discovery of new species is relatively common, rather than a rarity.

Bony fish and sharks have definite differences; the former have skeletons of bone while the latter use cartilage, the same tissue which forms human ears and coats the endings of bones as 'gristle'. Gills are used to obtain oxygen directly from water and, in deference to their highly active lifestyles, both groups require efficient means of moving. Sharks and the bony fish solved the problem in different ways, although both rely on fins. Sharks succeed using blunt force, swimming with muscle power and sinking if they rest. Fish, however, have developed swim bladders – air-filled sacks – for buoyancy, and can therefore effortlessly maintain a position in the water and use power to move only when required.

Birds and fish are only two of the five major vertebrate groups; we also have amphibians, reptiles and mammals. Here we have a succession in evolution: fish evolved to crawl from the water on modified fins in the Devonian period some 350 million years ago, breathing via primitive lungs which developed to gulp air while living in stagnant water. Such adaptations, arising by chance, permitted the beginnings of land conquest. Then came the early amphibians, slim, fish-like animals which developed legs based on fins (the supporting rays in the fish's lobed fins evolved into amphibian bones). While amphibians truly inhabited the land, they were forced to return to water for laying eggs. Today's salamanders, newts, frogs and toads are all confined in this way (some live for greater or lesser parts of their life in water), yet are diverse enough a group to possess over 4,000 species.

Amphibians are eye-catching in more ways than one, as their eyes are normally large and work well in both water and air, quite a trick considering the differences in refraction involved (the eye is deliberately distorted to permit correct focusing). They developed the means of detecting sound in air, and many possess poisonous skins, a fact made use of by South American natives who rub arrowheads in a frog's slimy coating. There is no point in having a well-developed defence system and not announcing the fact, so many creatures bear bright colours as a warning, in the same way that a security firm advertises its presence in the hope that trouble will go elsewhere.

With the coming of reptiles, the amphibian's dependence on water was superseded. With a dry, scaly skin rather than a moist, soft one, reptiles could enter a far greater range of habitats, from deserts to mountains, and lay eggs sealed against water loss. The dinosaurs are classed as reptiles (the word dinosaur means 'terrible lizard'), ruling the earth for 100 million years; in terms of time on earth it is a success story which far exceeds our own. Even though they are now extinct, their descendants live on as crocodiles, tortoises and lizards. Today's reptile species number 6,000.

Evolution, as it approached warm-bloodedness, seems to have experimented in a number of ways. With mammals came the ability to control temperature, develop young internally for live birth rather than using eggs, and suckle young using milk. With only 4,000 species, there is nevertheless a huge range of variation, though there are limits of size.

Small animals logically require less food than large ones. However, some of this food is used to generate heat. If the animal's body is warmer than its surroundings, it will continuously lose heat, which must be replaced: more food is required. Everything possible is done: fur becomes thicker, behaviour is altered to hibernate or migrate in cold periods, but still heat is lost. The problem lies with the area of skin: a small animal has a large surface area compared with its mass, or weight. In design terminology, there is a point where smaller sizes become untenable: the rate of heat loss exceeds the rate of supply, even if food is eaten continuously. This is the situation faced by some of the world's tiniest animals, shrews, which must eat at least every twenty minutes.

Heat loss therefore places a limit on the smallest size which warm-blooded mammals can attain, something which also has to take into account the size of young at birth. Some mammals – marsupials – evolved with pouches, to suckle relatively undeveloped young in a warm, protective situation. At its

With the advantages of warm blood came the need to maintain a constant temperature. A variety of methods are used. For example, animals which live in exceptionally cold conditions, such as penguins, and those with a low body weight, modify their behaviour to avoid rapid heat loss. Individuals in this colony of Long-fingered Bats, which inhabit a cave in Table Mountain, near Cape Town in South Africa, conserve their heat by grouping together in a mass.

smallest limit, *Planigale subtilissima* (a marsupial mouse) weighs about 10 g and grows to a maximum length of 95 mm. Even then, it is not the smallest mammal. The European Pygmy White-toothed Shrew (*Suncus etruscus*) can be as light as 2 g and measures only 35 mm long, while the adult Kitti's Hog-nosed Bat (*Craseonycteris thonglongyae*) from Thailand weighs only 1.7 g, with a body length of only 29 mm. At these warm-blooded sizes, survival would be impossible without a partial return to cold-blooded behaviour: the minute bat must pass into torpor at regular intervals.

At the other end of the scale, size is governed by movement. Bones are used for support, but are heavy and require muscles (which are also heavy) to operate joints. Larger animals require larger muscles, which add weight and require larger bones to support them, which require more muscle . . . Eventually, a point is reached where the bone is too immense to permit movement. It is no wonder that the largest animals on earth have lived in water, which relieves some of the burden of support. The dinosaur *Brachiosaurus* weighed over 50,000 kg, and *Diplodocus* attained a length of 30 m. The largest animal ever to exist is the Blue Whale, also listed at over 30 m long with a weight reaching 150,000 kg.

Form and function; there is always a reason for the shape and size of any animal or plant. While an over-riding feature of animals is that they eat ready-made food, plants manufacture sugars from raw ingredients during the process of photosynthesis. Plants originated in water as simple algae about 1,000 million years in the past; earth now possesses around 27,000 species. They gained complexity as seaweeds but, surrounded by the sea, there was never a need to develop roots or specialized mechanisms of obtaining water. With water in the soil, structures were required to obtain and transport it throughout the plant. Amongst the first plants to succeed was *Cooksonia*; fossils tell us that it possessed rudimentary 'roots' termed rhizoids, it could carry water internally and, as a bonus, it could use the same tissues to give it support and stand upright. Of crucial importance, the plant also had a coat which reduced water loss.

With this independence from the ocean, a vast range of plants could develop. The simplest are the mosses and their relatives (which include the liverworts), which now number 17,000 species. They survive only in damp conditions, and do not possess a true root, stem or leaf. Ferns (10,000 species) evolved leaves and fully functional vascular (transport) tissues, and paved the way for the gymnosperms (500 species). These coniferous trees bore seeds rather than spores, granting the advantage of protection and an enclosed food store to begin growth. Associations arose between different species, the most staggering involving the plant and fungal kingdoms: lichens, associations between fungi and algae, number 18,000 types.

All plants require sunlight energy to produce sugar. In air, this poses a problem as sunlight also produces heat. This restricts some groups to shady or damp areas, but the angiosperms – the flowering plants, well over 220,000 of them – evolved leaves which could cope with the trials and dangers offered by extreme temperatures.

To absorb sunlight, a leaf must be wide and, to avoid a wastage of cells, thin. This delicate structure must be cooled with a continuous flow of water to avoid damage, so a system of hollow cells connects the leaves to the roots, which take in water. As water evaporates, more is drawn up the stem by suction. The design, however, produces difficulties if water is in short supply, so there are mechanisms in the leaf to adjust the rate at which evaporation occurs: the leaf is a working factory, complete with moving parts and control systems. Before the supply of water reaches its predictable lowest, in the winter when it is frozen as ice and snow, the plant does what it must: it deposits its leaves on the ground, ejecting any unwanted chemicals in the process. This means of excretion gives us a blaze of colour as leaves are filled with metals and minerals before they are returned to the soil.

Flowering plants dominate the world and present an incredible display of biodiversity: riotous colours, and complicated, exotic designs are the norm. A flower is the reproductive centre of a plant. Its scent and colour are present to attract insects, supplying nectar as a bribe to ensure that pollen is carried to the next plant. Pollen is the male reproductive cell which must reach its destination: the female ovum, contained in the flower's ovary. Alternatively, pollen may be dispersed by the wind, a technique which wastes millions of airborne pollen particles, as testified by hay-fever sufferers worldwide.

Plants are the keystone of life on earth. Without their influence an oxygen atmosphere would not have developed, and there would be no food to supply the food chain: plant makes food, herbivore eats plant, carnivore eats herbivore. The inter-reliance of species is all the more remarkable when the plant/animal relationship is considered: without insects, flowers are a pointless waste of energy. Without flowers there can be no fruits to attract animals to disperse seeds. Plants release excess oxygen as waste, the very gas which animals consume; without plants the atmosphere cannot be maintained.

Life has colonized every corner of the planet. The polar ice caps maintain life which lies dormant until a riot of growth and a frenzy of feeding each spring. Deserts maintain lichens on bare rocks and sand, mostly desiccated during the day but becoming suddenly active at the least hint of moisture. Bacteria and plant spores coast the winds at over 1,500 m above the surface of the earth, while others survive the temperatures of super-heated springs.

Life is amazing. Its many forms stagger the imagination. It adapts, it evolves, it is resilient. We are fortunate to be able to revel in its diversity, all 1.4 million of its catalogued species, and to consider how much there is yet to discover in our natural, living world.

Mosses cover this fallen tree in the Adirondak Mountains of New York State. Without a true root, stem or leaf, mosses are considered primitive plants, even though they have successfully colonized the land.

C H A P T E R   F O U R

LIVING WITH NEIGHBOURS

Now do these complex and singular rules indicate that species have been endowed with sterility simply to prevent their becoming confused in nature? I think not. . . . Why, it may even be asked, has the production of hybrids been permitted? To grant to species the special power of producing hybrids, and then to stop their further propagation by different degrees of sterility, seems a strange arrangement.

Charles Darwin, *Origin of Species*, 1859

WHAT IS A SPECIES? Commonly, a species is defined using reproduction as the key element: a species can freely reproduce with others of the same kind, but not with other species. The ability to interbreed defines whether or not an individual is a member of that species.

As with so much else in the world, nothing is perfect and, while this concept of a species is the best that has been proposed, there are nevertheless huge inconsistencies. For example, some forms of life are considered separate, yet can interbreed. A female horse and a male donkey produce a mule, while a Brahman bull and Angus cow cross-breed to yield Brangus cattle. Similarly, a cow and a yak generate a dzho, and a grapefruit and tangerine form a mottled green and yellow fruit rather suitably called an ugli. Such offspring, created from breeding two dissimilar but related species (or well-marked varieties within one species), are called hybrids. The more closely related the two species are to begin with, the more possibility there is that they can interbreed.

That two species should exist which are so dissimilar that they are considered separate, yet close enough to each other that they can interbreed, is not surprising in the light of Darwin's theories. A new species has to arise from an older ancestor, and thus it will, during the early stages of its formation, be similar to its parental population: it would be more astonishing if these 'borderline' cases did not exist, as this would point towards a world where evolution had ceased. Evolution is not a theoretical process consigned to the distant past; it is active and around us every day.

Take a hypothetical example. A population of plants grows well in a region which has adequate rainfall, temperature and light, and is ideally adapted to its environment. Then a barrier arises: speed is not a factor, only that the original population is eventually divided. Perhaps land is split from the continental mass to form an island, mountains heave slowly into the skies and part the area into two, or polar ice caps melt and flood an intermediate, low-lying area. It was a feature of earth recognized by Alfred Wallace who, concurrently with Darwin,

Opposite: The Oxlip, *Primula elatior*, is an example of a hybrid formed from two dissimilar parents: a Primrose and a Cowslip.

developed a theory of evolution; the following text, published in 1855, pre-dated any public announcement of Darwin's theories:

> *The surface of the earth has undergone successive changes; land has sunk beneath the ocean, while fresh land has risen up from it; mountain chains have been elevated; islands have been formed into continents, and continents submerged till they have become islands; and these changes have taken place, not merely once, but perhaps hundreds, perhaps thousands of times.*

Although the plants in our hypothetical population, now divided by a barrier, are of the same species, individual specimens will have minor differences: this variation forms the keystone of natural selection. The separated populations will inevitably live under different conditions; perhaps the island habitat is moister or more exposed, or mountains now shield their leeward side from excess rain. Under a different environment, evolution *must* occur and the plant populations will travel their own paths. If the change of habitat is too extreme, the plants die. If the differences are slight, evolution will be slow – but it will occur.

In time, perhaps the two populations are reunited; the land is rejoined, or the seas recede. There may be clear, physical differences between the populations, representing the beginnings of speciation. However, if evolution has not been permitted enough time, or environmental pressures have not been great enough to cause change, then they will freely interbreed and exchange genes: the gene pool is remixed and a single species remains. If, on the other hand, the plants have diverged far enough, they cannot produce any progeny: a new species has formed. There are therefore two elements involved in the evolution of new species: a separation of genes and sufficient time.

The barrier which formed between the populations is termed an isolating mechanism. It may be a geographic one, such as mountains or water, or it may be behavioural: all that is necessary is that two sections of a population are subjected to an effective obstruction which prevents the exchange of genes. Different plants, related closely enough to produce hybrids, may grow in the same area but mature at different times. As it is never possible for one to pollinate the other they are effectively isolated and never the twain shall (biologically) meet – even when two plants grow with roots intertwined, or animals share the same grazing area. Evolution can proceed and they may grow further apart until interbreeding is no longer an option. The mechanisms involved have accentuated the probability that new species will be produced.

There are many examples of new species in the making, giving real substance to Darwin's theories. Spoil heaps in North Wales, the abandoned remnants of mining, contain what would normally be lethal cocktails of copper, lead and zinc. However, a few specimens of the Common Bent Grass, *Agrostis tenuis*, have been found growing on the edge of the heaps; the original population possessed enough variation for some individual plants to survive the metallic poisons, while the rest were eliminated. Over time, these tolerant grasses interbred and reinforced their differences; a 'normal' seed would not grow and only the 'new' variants survived to breed again.

The bent grass demonstrates the effects of an environmental change. Grasshoppers belonging to the *Chorthippus* genus, *C. brunneus* and *C. biguttulus*, illustrate isolation within a population. The two species appear almost identical but have different mating calls. When the male stridulates (makes its species-unique chirruping sound) it triggers mating only in the 'correct' female. However, if a tape recording of the 'correct' type is played to a female, she mates with any available male of the 'wrong' species and produces a fertile hybrid. Reproduction between the species is possible, but behaviour maintains a separation.

Just as the tolerance for waste ground increases in the bent grass as it moves towards being classified as a new species, the grasshoppers are diverging slowly and eventually will no longer be able to interbreed. This is presumed to be occurring in two species of British hawthorn: Common Hawthorn (*Crataegus monogyna*) and Midland Hawthorn (*C. laevigata*). Common Hawthorn grows on open ground, while Midland Hawthorn grows in dense woodland; the leaf shapes differ considerably. The two plants can interbreed to produce fertile hybrids (with an intermediate-shaped leaf), but their distinctive leaves are sufficiently different to consider them as separate species.

A perfect example exists with two species of bird found in the northern hemisphere: the Herring Gull and the Lesser Black-backed Gull. These occur in a series of sub-species, each of which is a slightly different colour from the others. What is of interest is the degree of possible interbreeding.

The ancestral population of gulls is believed to have lived in Siberia, an area now occupied by the Vega Gull (*Larus argentatus vegae*). This can breed with the herring gull sub-species on either side: *L. argentatus smithsonianus* in North America and *L. argentatus birulaii* in Scandinavia. The Scandinavian gull can also breed with *its* neighbour, a black-backed gull: *L. fuscus heuglini*. In so doing, the herring gull has jumped species to the black-backed gull, after which the smoothly graded genetic chain continues, through adjacent sub-species populations, until it reaches Britain and the Lesser Black-backed Gull, *L. fuscus graelsii*. Interbreeding has circled the pole, but the two ends of the arc cannot interbreed: the American Herring Gull and the Lesser Black-backed Gull deserve their species' status.

Such situations illustrate the difficulties in assigning a species name to a population, and the folly of insisting too rigorously on a definitive sub-species. Any of these groups can be considered to be evolving, on their way to full species. Whether the adjoining black-backed and herring gull sub-species populations are assigned correctly to their nearest relatives is irrelevant: they could be placed with either. Darwin himself noted the problem of defining species and variations within a species when he wrote:

> *Certainly no clear line of demarcation has as yet been drawn between species and sub-species – that is, the forms which in the opinion of some naturalists come very near to, but do not arrive at, the rank of species: or, again, between sub-species and well-marked varieties, or between lesser varieties and individual differences. These differences blend into each other by an insensible series.*

Isolation mechanisms, leading to the formation of what we determine to be new species, are inherently slow. However, a natural and rapid means of speciation also exists. To explain how the system operates, an understanding of how information is passed from one generation to the next is required.

From the beginning of our natural world, life has been based on DNA, which forms the essential part of a chromosome. DNA forms a blueprint; it contains a set of data, called genes, which control every aspect of the cell's existence. What is particularly important to the cell is the production of protein, each type having different properties. Proteins are made of long chains of amino acids, so what is required is the information to link these acids in the correct sequence. Each piece of data – each gene, individually or as part of a group – controls a facet of the organism such as the colour of eyes, hair or skin, or the formation of different cells such as muscle and nerve. Genes, then, are at the basis of life: they are in effect at an organizational level beyond the normal series of classification: species, individual organism of that species, gene.

It should be immediately obvious that the genetic data in the chromosome must be extensive. To carry all the information, about three metres of DNA is required in the human species. If this measurement was scaled up to make DNA the thickness of string, it would produce an unravelled length of about 1,000 km. This sort of structure would be unmanageable by the cell, but two factors aid its packaging within the nucleus. First, the total length is divided into separate strands which form chromosomes, and these are coiled and

THE DNA LADDER OF LIFE

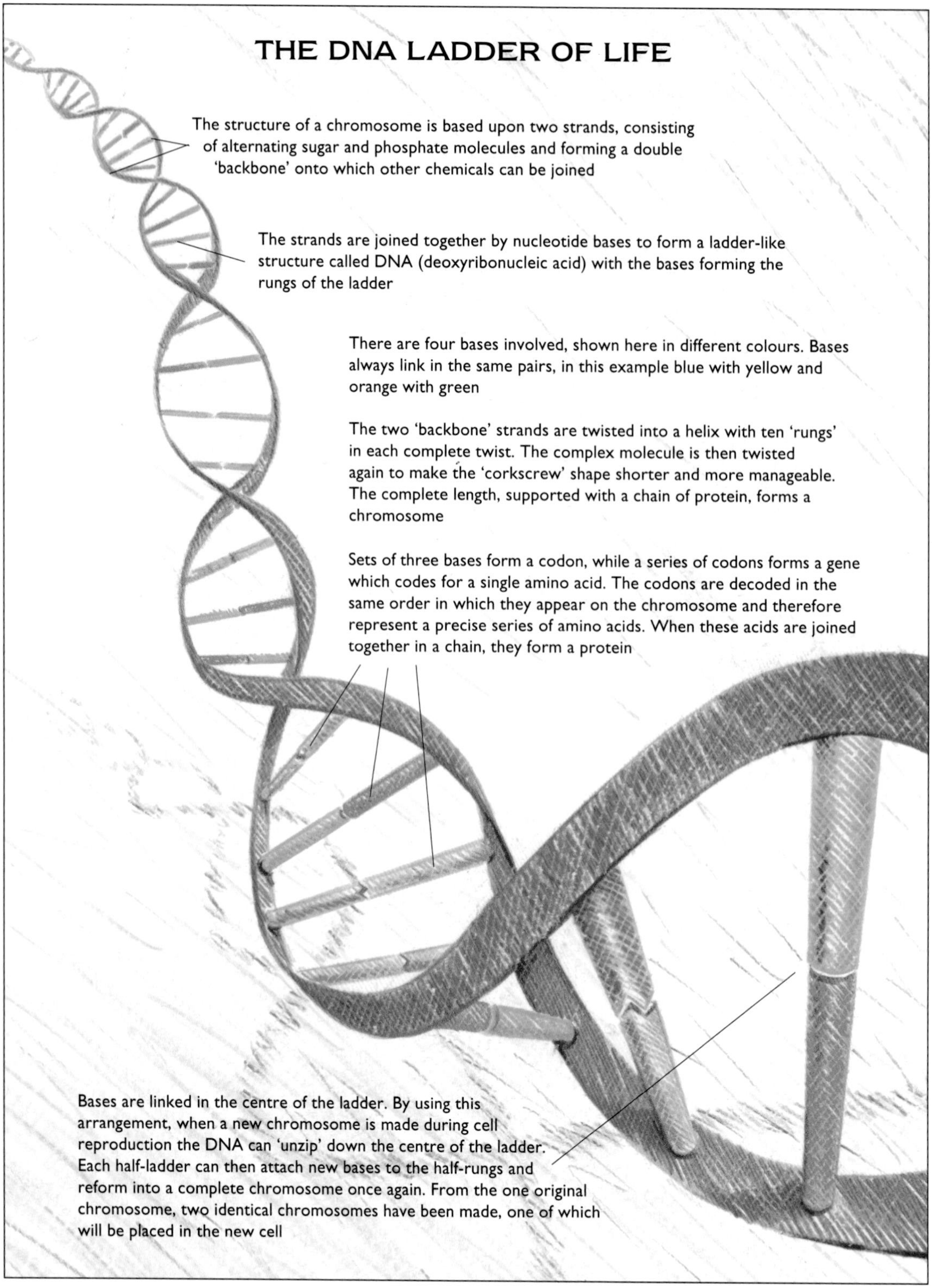

The structure of a chromosome is based upon two strands, consisting of alternating sugar and phosphate molecules and forming a double 'backbone' onto which other chemicals can be joined

The strands are joined together by nucleotide bases to form a ladder-like structure called DNA (deoxyribonucleic acid) with the bases forming the rungs of the ladder

There are four bases involved, shown here in different colours. Bases always link in the same pairs, in this example blue with yellow and orange with green

The two 'backbone' strands are twisted into a helix with ten 'rungs' in each complete twist. The complex molecule is then twisted again to make the 'corkscrew' shape shorter and more manageable. The complete length, supported with a chain of protein, forms a chromosome

Sets of three bases form a codon, while a series of codons forms a gene which codes for a single amino acid. The codons are decoded in the same order in which they appear on the chromosome and therefore represent a precise series of amino acids. When these acids are joined together in a chain, they form a protein

Bases are linked in the centre of the ladder. By using this arrangement, when a new chromosome is made during cell reproduction the DNA can 'unzip' down the centre of the ladder. Each half-ladder can then attach new bases to the half-rungs and reform into a complete chromosome once again. From the one original chromosome, two identical chromosomes have been made, one of which will be placed in the new cell repeatedly twisted again to make the chromosome thicker but shorter and more controllable.

The chemical structure of DNA relies upon two long chains of alternating sugar and phosphate molecules. Each unit of sugar is linked to its counterpart in the other chain by a pair of nucleotide bases (usually simply referred to as 'bases'), like the rungs of a ladder. The complete unit of DNA is then twisted into what is termed a double helix (because the shape is like a coiled spring, with two intertwined strands supplied by the two 'backbone' sugar–phosphate chains). If this shape is hard to imagine, think of it as a spiral staircase where the two sides are DNA strands and the treads form the bases: the whole staircase is a chromosome.

By using bases in pairs, joined with each other in the centre of each rung, DNA can 'unzip' down its centre during cell reproduction. Each half-ladder then forms a template which is used to reconstruct the 'missing' half-ladder and hence produce two new, complete strands of DNA. In this manner, chromosomes are replicated to supply the new cell.

The bases forming the rungs of the ladder can be any one of four types: adenine, cytosine, guanine or thymine. Now, imagine a box of children's building blocks made in four colours. A set of three blocks (bases) attached to each other comprises a single unit of information which represents a specific amino acid, and which is called a codon: a blue–blue–blue sequence is a codon which represents one amino acid, while blue–green–blue represents another. In the spiral staircase analogy, each successive set of three treads forms a single piece of data. In the chromosome, the codons are fixed in a sequence and are 'read' by the cell in the order in which they appear. The complete set of codons carries the information for a single protein, the sequence being termed a gene. The cellular dictionary decodes the genes, and links the corresponding amino acids together in the same order. The order of the codons within the gene therefore controls the sequence in which amino acids are attached to the protein chain, just like reading a book from the start to the end. The genes appear in sequence on the chromosome, like an encyclopedia of books on a shelf.

Altogether, each protein might be made up of 500 amino acids (each requiring a gene to code for it), and would inherently require a specific sequence of up to 2,000 bases. To code for all the different proteins needed by the organism there are about 1,000 genes in bacteria, and upwards of some 400,000 genes in animals and plants. Human chromosomes carry about 100,000 genes.

Chromosomes normally occur in pairs. Their total number does not appear to be closely linked with the complexity of the organism, although extremely simple cells such as bacteria generally have fewer chromosomes than, say, mammals or birds. Man, for example, has 23 pairs of chromosomes in each cell; crayfish have 100 pairs, some ferns have 250 pairs, dogs have 39 pairs, mice have 20 pairs, rye 7 pairs, while fruit flies have 4 pairs.

The whole process of reading and reproducing the sequence of bases and genes is like a computer coding which runs a massive program: if there is any one piece of coding that is incorrect, the program may not run, it may run perfectly until attacked by a virus (a harmful piece of computer software akin to a biological virus), or it may exhibit 'bugs' (unexpected glitches which the computer user has to contend with). Computers, with 'bugs' and 'viruses', possess many analogies to life, but there are limitations with this parallel. Genes have a physical form; the information they carry is more akin to the punched cards of early computers than to transient electronic data that appears on a screen. In this analogy, the card is DNA while the punched instructions (in groups upon the card, performing different tasks as part of the whole) are genes. A number of cards are required to complete the programme's operation, just as a number of chromosomes are found within each cell.

In evolutionary terms, if the computer program, the genetic code, is poor it is discarded (the species becomes extinct, or the individual dies) or it is modified (a new species forms; the

PASSING ON THE CODE

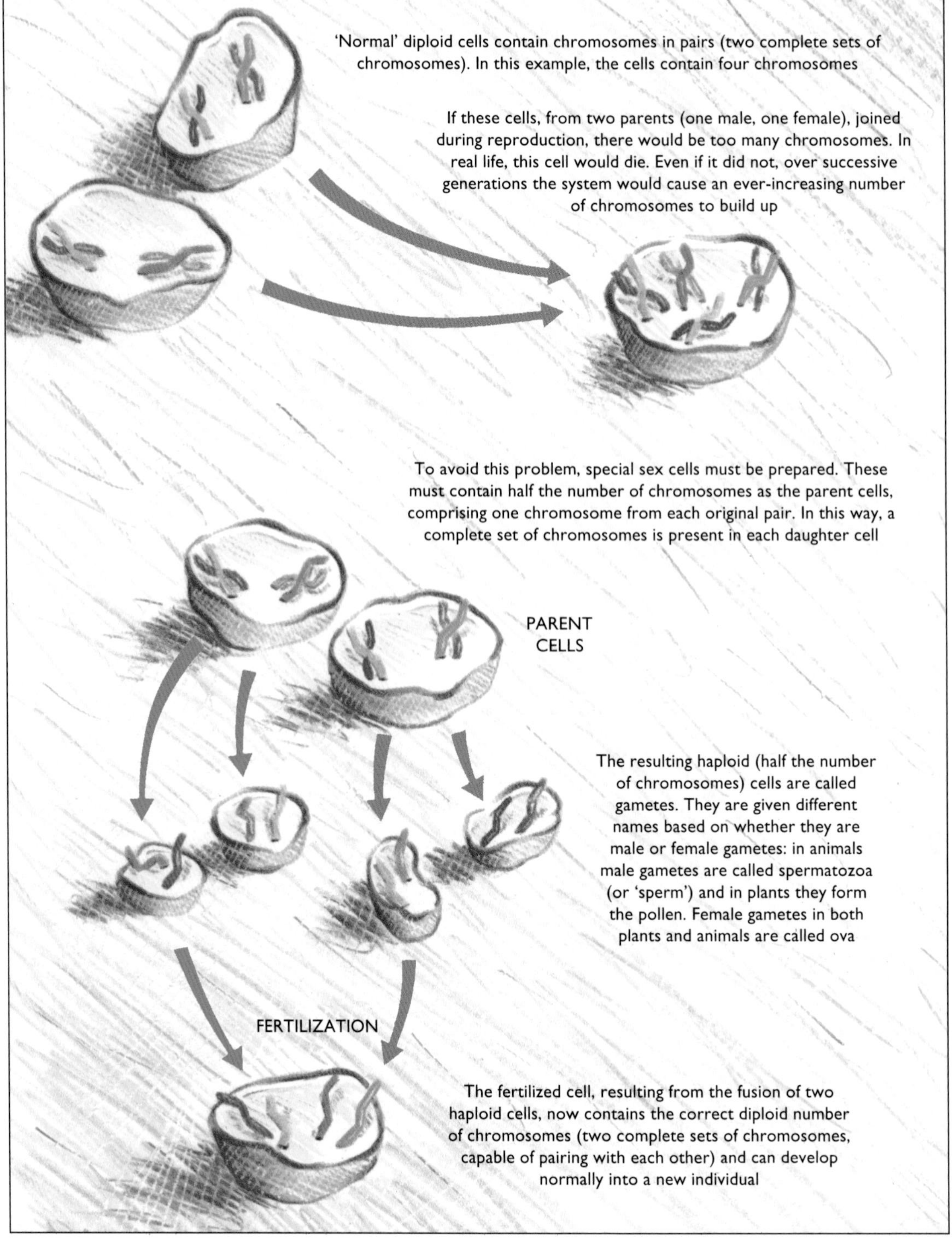

'Normal' diploid cells contain chromosomes in pairs (two complete sets of chromosomes). In this example, the cells contain four chromosomes

If these cells, from two parents (one male, one female), joined during reproduction, there would be too many chromosomes. In real life, this cell would die. Even if it did not, over successive generations the system would cause an ever-increasing number of chromosomes to build up

To avoid this problem, special sex cells must be prepared. These must contain half the number of chromosomes as the parent cells, comprising one chromosome from each original pair. In this way, a complete set of chromosomes is present in each daughter cell

PARENT CELLS

The resulting haploid (half the number of chromosomes) cells are called gametes. They are given different names based on whether they are male or female gametes: in animals male gametes are called spermatozoa (or 'sperm') and in plants they form the pollen. Female gametes in both plants and animals are called ova

FERTILIZATION

The fertilized cell, resulting from the fusion of two haploid cells, now contains the correct diploid number of chromosomes (two complete sets of chromosomes, capable of pairing with each other) and can develop normally into a new individual new version of software may not understand – breed with – its old form). In the cell, the wrong amino acid may inadvertently be used, or be inserted the wrong way round. Faults and mistakes arise. Errors can, and frequently do, occur.

The manufacture of sex cells and the chromosomes they contain is of crucial importance. Cells with the full complement of paired chromosomes are called diploid cells. When sexual reproduction takes place, a cell from each parent fuses with its partner, introducing a new mixture of genes for the offspring. This presents a problem: the new cell would have twice as many chromosomes as it should. This is the reason for the presence of the specialized sex cells (or gametes): pollen and ova in plants, sperm and ova in animals. These all bear only one of the original paired set of chromosomes and, as they now have half the parental number of chromosomes, they are termed haploid cells. When reproduction takes place, two of these haploid cells can join together to make an embryo cell with a full diploid (two-set) content.

During the gamete's production stage, an error can occur which is crucial to the offspring. Occasionally, instead of dividing into two chromosome sets, an extra chromosome is placed in one sex cell, leaving another sex cell with one chromosome too few. Down's syndrome, formerly known as mongolism, results from the presence of one extra chromosome in the ensuing fertilized cell. Turner's syndrome, which produces sterile females, is due to the absence of one of the chromosomes controlling sexual characteristics. It has been suggested that an extra male-controlling 'Y' chromosome in men increases psychological disorders; statistically, criminals have a higher percentage of this trait than the general population.

Of more importance in the production of new species, sometimes a sex cell forms which contains *all* the parent's chromosomes: a diploid gamete. Now, when reproduction occurs, and this diploid gamete fertilizes a normal haploid gamete, the result will be a cell with three sets of chromosomes: a triploid. In general, cells which contain a greater number of chromosome sets than they should are termed polyploids: this can be any number based on multiples of the haploid chromosomes.

The new polyploid animal or plant which arises is genetically very different from its parents: it can no longer breed with the population of organisms which produced it. On the face of it, in biological terminology, a new species has been formed. Of even more interest is the speed at which this has been done; there are no interminable years to wait while variations are subjected to the forces of nature and natural selection runs its course.

However, the new polyploid organism has problems. It not only contains an 'incorrect' number of chromosomes, these may no longer be compatible. When sex cells are formed, chromosomes first pair off to ensure that a complete set is placed in each new cell. When there are uneven sets of chromosomes this is not possible; it appears that the chromosomes become tangled together, resulting in either a totally abnormal cell which dies, or total failure of the process. Without the ability to manufacture reproductive cells, the organism is infertile and cannot even breed with its own type.

The same result – an imbalance in chromosomes leading to sterility – applies when two related species interbreed to produce hybrid (rather than polyploid) offspring. The product of a horse and donkey, a mule, is sterile even though it has a normal sex drive. Horses have 32 pairs of chromosomes, while donkeys have 31 pairs, causing the resulting embryo cell to receive an uneven total of 63 chromosomes. Here, the reason for sterility is slightly different, though still linked with chromosome imbalance. In order to produce sex cells the chromosomes must pair off perfectly (comparable chromosomes must lie alongside each other during the process of cell production, as some genes are exchanged between them; if the line-up is not perfect, the 'information exchange' is scrambled). If the two chromosomes in the pair are not sufficiently alike, the process fails and no sex cells are formed. The hybrid offspring, although possibly able to develop normally, are sterile. Hybrid animals and plants are therefore those which result

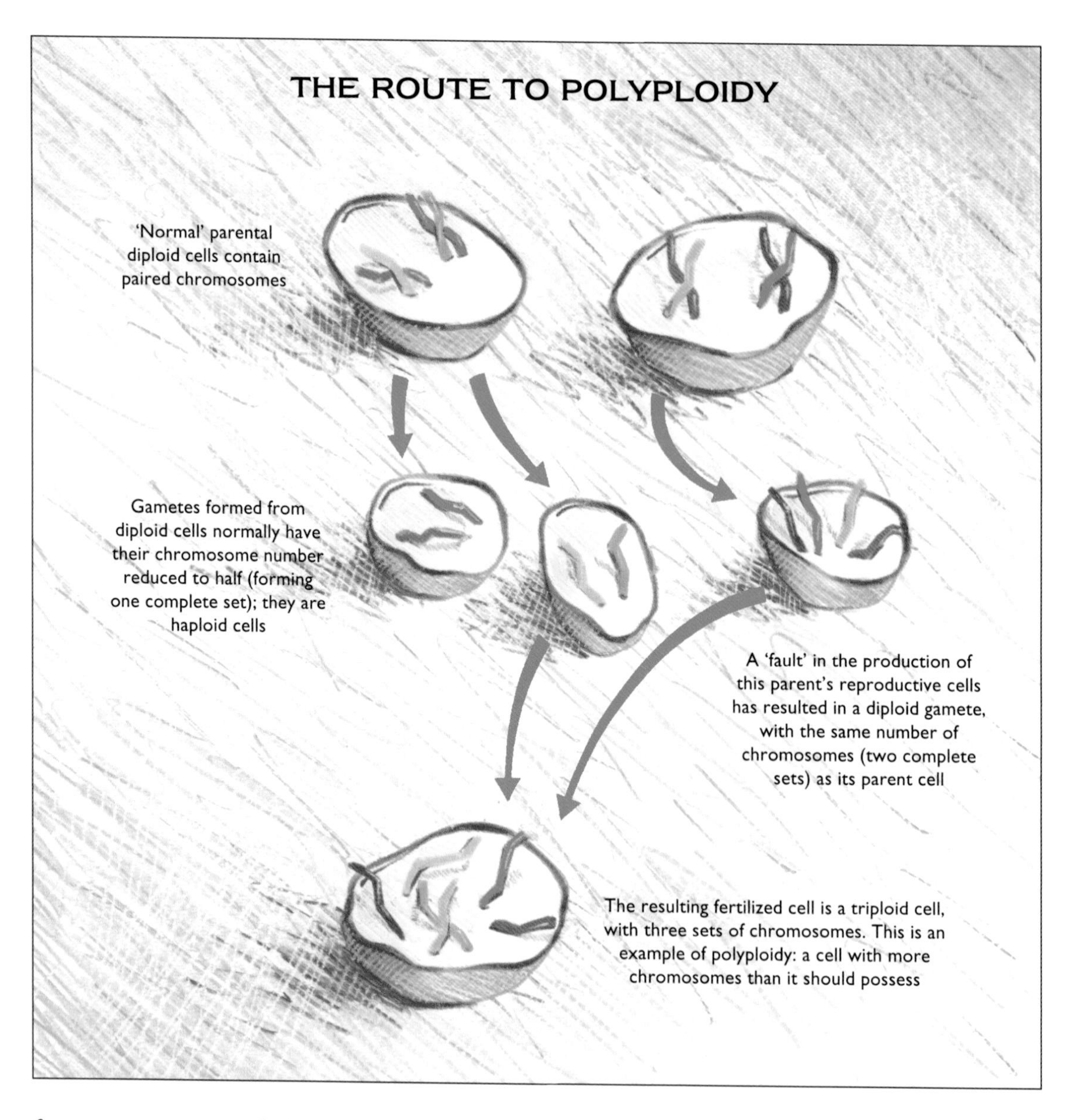

from two parents of different but related species, rather than direct changes in the chromosome number of an individual organism which produces a polyploid.

The chance of hybrid sterility is increased with a greater separation between the two species (because the chromosomes are more dissimilar), but most hybrids are sterile in any case. In animals, sterility is genetically catastrophic as, without the capability to breed, there can be no continuation of the species into the future. There may be a payback, though, as many hybrids combine the strengths of each of their parents without the corresponding weaknesses: mules, for example, are stronger than horses or donkeys.

This advantage, termed hybrid vigour, also applies to plants. Sterility in plants is not always such a catastrophe, as an alternative form of reproduction may be available. 'Daughter' plants can be produced via structures such as bulbs (for example, daffodils),

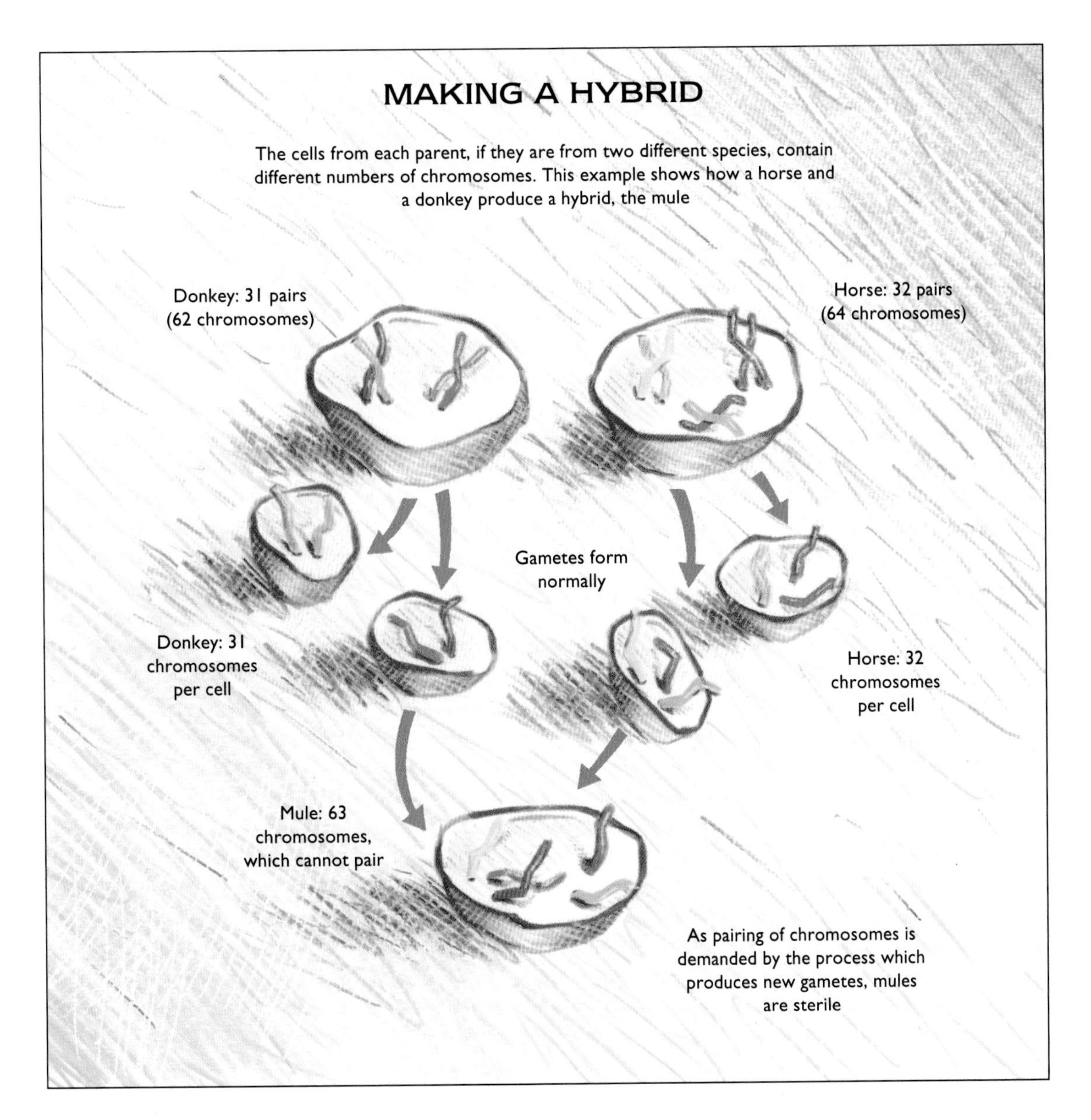

tubers (dahlias) or runners (strawberries). Reproduction in this manner, referred to as asexual (one parent) reproduction, offers the capability of increasing the number of individual plants, although each new generation will remain genetically identical to its parents, as well as being sterile.

Sometimes, this ability to combine the means of rapidly and easily increasing in numbers with a vigorous, advantageous set of characteristics, is enough of an advantage to a new species of plant that it may displace its parents. Cord-grass (also known as rice-grass), which grows in salt marshes, is a prime example.

A variety of cord-grass, *Spartina maritima* (sometimes referred to as Common Cord-grass, though it is now comparatively rare), was once prolific in the marshes of south-eastern Britain. During the 1800s the related *S. alterniflora* (Smooth Cord-grass) was accidentally brought by ship to Britain from North America, and a population established itself near Southampton.

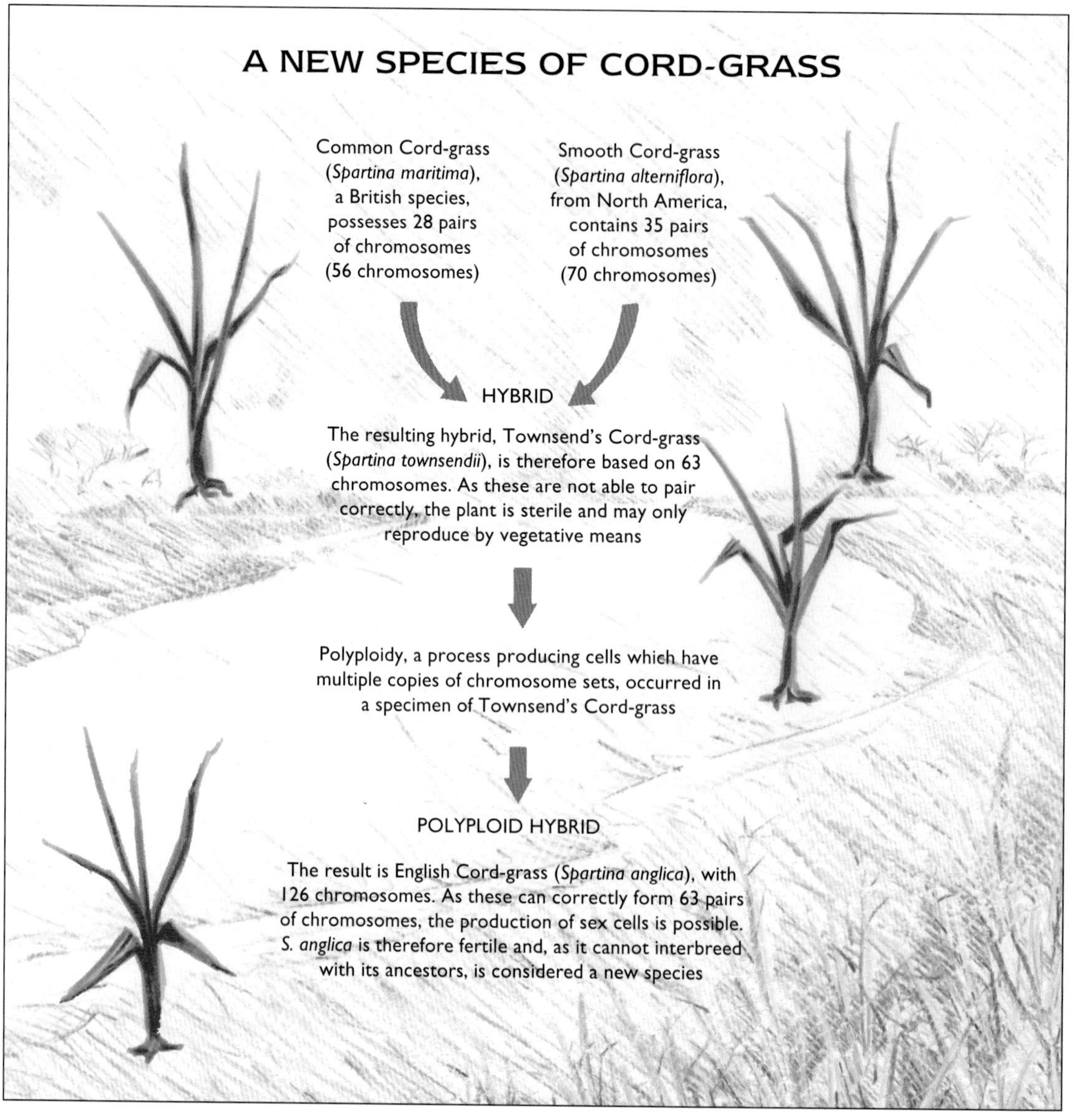

A NEW SPECIES OF CORD-GRASS

Common Cord-grass (*Spartina maritima*), a British species, possesses 28 pairs of chromosomes (56 chromosomes)

Smooth Cord-grass (*Spartina alterniflora*), from North America, contains 35 pairs of chromosomes (70 chromosomes)

HYBRID

The resulting hybrid, Townsend's Cord-grass (*Spartina townsendii*), is therefore based on 63 chromosomes. As these are not able to pair correctly, the plant is sterile and may only reproduce by vegetative means

Polyploidy, a process producing cells which have multiple copies of chromosome sets, occurred in a specimen of Townsend's Cord-grass

POLYPLOID HYBRID

The result is English Cord-grass (*Spartina anglica*), with 126 chromosomes. As these can correctly form 63 pairs of chromosomes, the production of sex cells is possible. *S. anglica* is therefore fertile and, as it cannot interbreed with its ancestors, is considered a new species

In 1870 a new hybrid species of grass, *S. townsendii* (Townsend's Cord-grass) was discovered. It was capable of rapid growth, and quickly spread using stolons (horizontally growing stems), forcing out the native species wherever the hybrid and its ancestral parent came into competition. Today, the hybrid plant is the dominant one in all southern marshes and is spreading slowly around the British coastline. However, as is usual in these cases, Townsend's Cord-grass is sterile.

As with the mule, an unbalanced set of chromosomes had been produced: *S. maritima* contains 28 pairs, *S. alterniflora* 35 pairs, resulting in *S. townsendii* with 63 chromosomes and unable to reproduce. Here, though, is one of the amazing areas of evolution: the instant production of a new species.

During some early stage of growth, polyploidy affected the developing cells of a Townsend's Cord-grass plant. Its 63 chromosomes were doubled to 126, and it suddenly

Left undisturbed, bluebells are able to propagate using their underground food storage organ, the bulb, until huge areas of forest are covered. Any new plants arising by this asexual process are clones, and are genetically identical to their parent.

possessed a balanced number of chromosomes, each able to pair off. As such, the new plant – English Cord-grass, *S. anglica* – was fertile. This polyploid hybrid species is now grown alongside its sterile relative to help reclaim mudflats. As English Cord-grass cannot breed with its ancestors (even the fertile ones), it is classed as a true, genetically different species.

Where hybrids and polyploidy are involved, therefore, a new species can rapidly and effectively arise, and over half of all flowering plant species exist due to these sudden polyploidal changes. As well as cord-grass, swedes are known to be polyploid plants produced from hybrids between a cabbage and a turnip, and breeding a radish and cabbage together forms a sterile hybrid which, if polyploidy occurs, results in a fertile 'cabbish'. Common Hempnettle (*Galeopsis tetrahit*), Wild Pear (*Pyrus communis*), Dog Rose (*Rosa canina*) and Sun Spurge (*Euphorbia helioscopia*) are also all polyploidal hybrids. Most potatoes are polyploids, based on a haploid set of 12 chromosomes: species may possess 24, 36, 48, 60 or 72 chromosomes. Wild potatoes are usually diploid (24 chromosomes), while most cultivated potatoes carry 48 chromosomes – and are sterile.

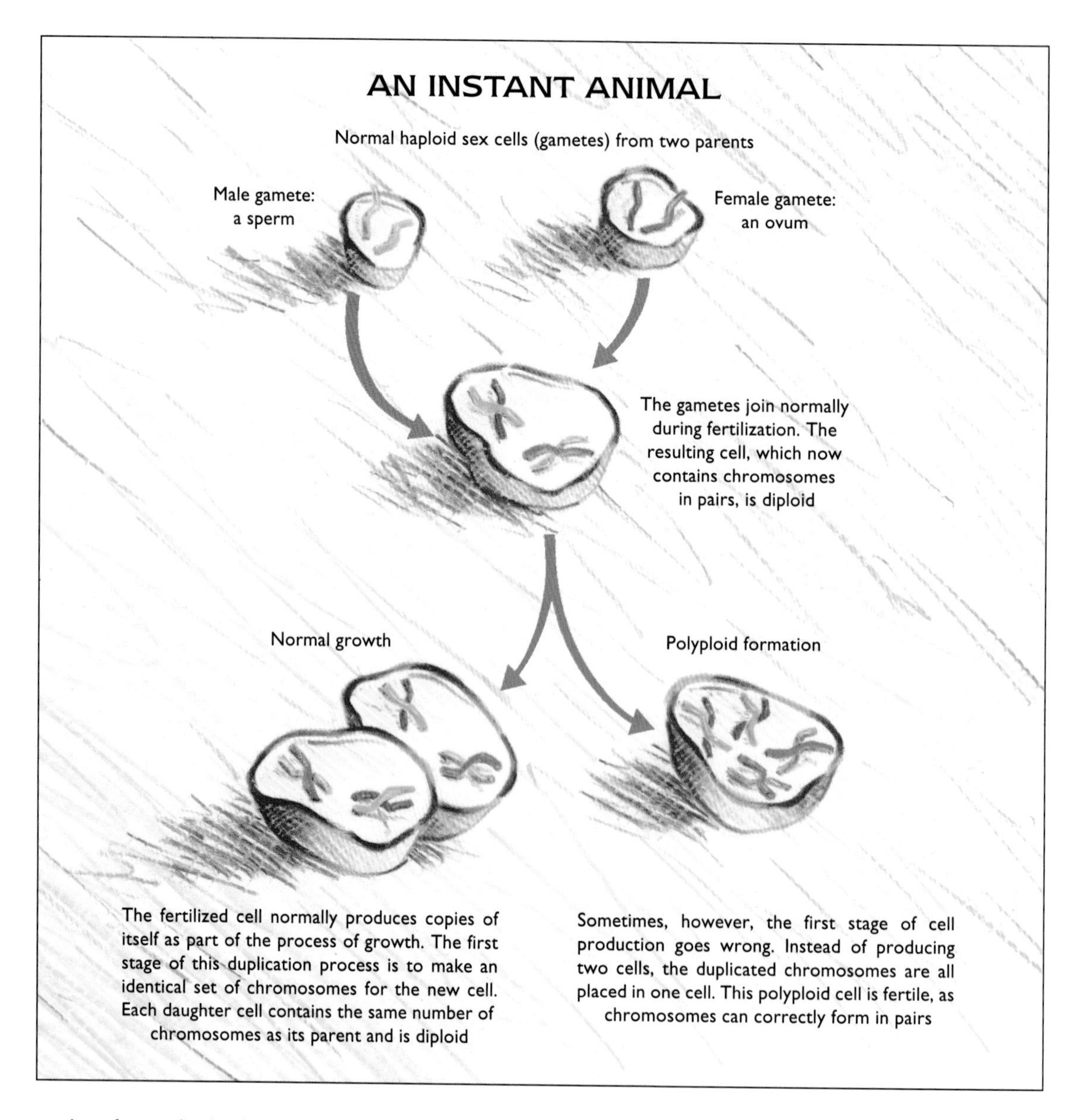

Another polyploid situation can arise which, although it can also apply to plants, is of particular significance to animals. As was intimated earlier, polyploidy is an unlikely route for animals to take towards speciation; they do not have the advantage of forming new individuals by vegetative means, so sterility indicates an immediate end to the new species. However, fertile polyploid organisms can develop. Even if the organism is fertile, though, it cannot reproduce without a mate, so polyploid animal species remain rare. For some reason, a high percentage of frogs are polyploids with double the number of chromosomes found in closely related species. The link between polyploids and amphibian species may be due to the occasional capability of unfertilized frog's eggs to develop into young, effectively causing an increase in the population of polyploid animals, which can then breed.

This road to speciation begins with the normal diploid cell formed during fertilization.

In order to grow into an organism, the single cell must make copies of itself. In this process the chromosomes are first duplicated so that a full set may be placed in the new cell. If, at this early stage, the process is faulty, it is possible for a new cell to be made which contains all the chromosomes, duplicates and originals; as the diploid cell contained chromosomes in pairs, this polyploid cell therefore contains chromosomes in groups of four. As these may correctly pair off, the four-set cell is fertile.

The objective of man's deliberate attempts to produce hybrids is to combine the 'best' characteristics of each parent strain or species, such as when producing a new colour of rose or type of rice. The techniques involved in this artificial selection process are simple, and known to every gardener and animal breeder. Selecting and mating specimens, then making a further selection from any ensuing offspring before continuing the process, may by chance produce a new variant. It may not, of course: such matters lie with the combination of genes.

It is therefore an accelerated process of trial and error which, commonly, ignores other factors and may seriously weaken the gene pool as, in striving to get the line to 'breed true', efforts are made to remove unwanted genes as they have an uncanny ability to re-exhibit their characteristics in later generations. Typically, the first generation of offspring may show the required characteristics, but when these are bred a proportion of the results are not as expected; perhaps a population of red flowers which interbreeds suddenly exhibits some white ones. In this case, according to the laws of heredity, only one gene of the pair was used by the organism, while the other remained dormant but available to future generations; breeders may find it difficult to eradicate such extraneous genes.

Examples of the successes and difficulties in controlling nature are rife, but the technique is ancient. Arabian horses date back to breeding programmes in about 500 BC and, during his period of rule over the kingdom of Prussia from 1740 to 1786, Frederick the Great endeavoured to increase agricultural production by introducing new, scientific methods of cattle breeding. Even humans have been selectively bred to increase the prevalence of certain characteristics. In 324 BC, well before the time of Hitler's proposed master race, Alexander the Great, King of Macedonia, introduced a breeding programme to create a body of nobles with mixed blood, using his own country's male aristocrats and Iranian women. In more modern times, Brangus cattle were bred in a bid to retain the characteristics of high-quality Angus beef with the better resistance to disease and pests of the Brahmen, but it took 30 years of breeding before stray genes were eliminated and the line was designated as purebred, producing the Santa Gertrudis breed.

This is one example of attempts to speed up what could theoretically occur naturally; it is possible for strains of plants or animals to interbreed randomly and produce the same results that man engineers, but it would take far longer to attain if chance was left as the deciding factor. By selecting the characteristics which man desires, then arranging for selective breeding, evolution is given a directed, controlled avenue to move along. It is surprising how many modern plants and animals which we take for granted have been bred for a purpose. Pigs, chickens, and animals or plants where there are recognized pedigrees, breeds or strains (dogs, cats, cattle, roses, marigolds . . .), all fall into this category.

Corn (maize) is a type of grass which produces a rich head of seeds; if, the theory goes, more and larger seeds can be produced with a hybrid, more money can be made with each field of crop. A greater yield of food (not to mention financial gain) is possible. Amongst the variants now being grown are dent corn (which contains a high percentage of starch), waxy corn (with a waxy coating, used to produce an alternative to tapioca), sweet corn, flour corn, and popcorn (popcorn's strong seed case resists the internal pressure exerted by steam as it is heated, until it explodes). Flint corn and podcorn varieties are probably the closest to the original variety brought back to Europe by Columbus, from which the other hybrids were developed. Hybrids have certainly been successful, not only in increasing yield by sometimes hundreds of per cent, but also by introducing new flavours and textures to our dinner table. When you consider the unlikely benefits to a plant of producing square tomatoes which leave less empty space when packed, seedless varieties of grapes and oranges, and a cow's milk production taken into the realm of ten gallons a day, you have to assume that man's hand has been present.

The crocus, introduced to Europe from Asia, has a long history of use. The dried, yellow stigma of *Crocus sativus* is used as a spice in the form of saffron. The Meadow Saffron, *Colchicum autumnale*, looks similar but is unrelated, and yields an alkaloid drug called colchicine which is used to treat gout. Colchicine is also used by plant breeders to encourage the formation of polyploid plants, in the hope of increasing their vigour. Over half of all flowering plants are polyploids.

The key to biodiversity therefore lies with genes. Genes are at the centre of the production of new species, whether by natural or manipulated means. When the genetic code is altered so that the population can no longer interbreed with its predecessors, a new species has been formed. However, as we have seen, evolution depends on the existence of variation in the population's characteristics for natural selection to act upon. In turn, this requires a variety of genetic material to select from.

Genes may act individually, or in combination. Some characteristics, such as the ability to roll your tongue and the wrinkled or smooth appearance of peas (one of Mendel's experimental factors), are under the control of a single gene, but others require a bank of genes to act in concert. In either case, the population contains different forms of the same gene which control one characteristic. It is akin to having a bag containing, say, 100 marbles. Many are identical, but others will be different sizes, colours and patterns. You are permitted to take out two, blindfolded, and must then act upon the random choices you have received. You might draw out two the same, or two different; these represent the genes you acquire, on a random basis, one from each of your parents, and which control a single characteristic.

Human blood groups are controlled in this way, with three different genes (marbles) available in what is referred to as the gene pool (marble bag). The gene pool is actually defined as the total available genes of all types for all characteristics in the population, while this example is restricted to just one characteristic. The argument nevertheless remains the same: each characteristic may be defined by one or many genes, and each gene may exist in a single or many forms. Each form of the same gene is termed an allele, of which there may be different proportions in the gene pool. If there are ninety-nine red marbles in every hundred in the marble bag, the chances are high that two red marbles will be received by random choice. If a red marble equates to a red flower, another colour may be a rare event. In human terms, individuals in minority ethnic groups are outnumbered by other populations, and the genes controlling their traits are therefore present in the human gene pool in low proportions.

The concept of the proportion of alleles within a gene pool is important. When a new species arises as a result of hybridization or polyploidy, there is only a severely limited part of the original gene pool in the new population; if the species has arisen due to artificial selection, alleles may have been deliberately or inadvertently removed. Added to this, the relative numbers of any allele will fluctuate from year to year, generation to generation, as individuals receive or pass on their genes, and are born or die. Any one of those genes, at any point in time, will possibly be in low ebb in proportion to its alternate alleles. That particular gene might be crucial to the future viability of a small population, especially when the gene pool is limited, as it is in hybrid populations.

Consider the laws of chance when a coin is flipped in the air: it has an even chance of coming down heads or tails. If only a few throws are made the results will probably not conform to this 50:50 ratio; a larger number of flips are required. In the same way, if there are insufficient individuals in the population to carry all the options, the chances of a particular allele surviving are affected. In a large population chance fluctuations are readily masked, and variations – and the ability to adapt – are preserved for the future: there are a large number of coin flips and proportions are maintained. However, in a small population (as with a small number of coin throws) the chance of alleles being preserved is decreased (the coin-flip probabilities are no longer 50:50) and there is 'genetic drift'. Below a certain level of population, alleles (in particular the less common ones) are invariably lost due to fluctuations in their frequency and the fact that there are no longer enough individuals to carry them.

Genetic drift is therefore vital in conservation terms. Animals and plants which are brought to the edge of extinction inevitably possess a decreased range of genetic material in their gene pools, and therefore will not exhibit the same variation in future generations. Their ability to adapt to changes in the environment is diminished, and the rate and efficiency of blind evolution has been curtailed; there is less chance of adapting to changing environmental conditions.

Realistically, when populations fall below about 500 individuals they can be assumed to have lost unique genetic material and are in risk of species failure, even if the population subsequently rises. Computer models show that fifty individuals have about a 95 per cent chance of survival as a species in the wild, but a natural catastrophe can easily tip the balance into extinction. In 1989 a hurricane struck the island home of the Puerto Rican parrot, *Amazona vittata*, already existing in a reduced population of about sixty birds. Some parrots had their feathers blown off, or were carried out to sea: others became easy prey for hawks, as their treetop concealment was stripped of leaves. The result was a remaining population of some thirty parrots. It has yet to be seen if the parrot, reduced below the simulation's 95 per cent level, survives in the long term, but it is clear that maintaining the gene pool by conserving a viable number of individuals is essential in a changing world, where genetic disaster may arise due to an unexpected misfortune.

The elements of isolation, time and space also affect the evolution and survival of a species. When a species is isolated in a stable environment, and given enough time, it will adapt ever more precisely to its specific living conditions. This isolation also implies a small population, so genetic drift remains important. When there are restrictions on a population's movement, whatever its nature, the living area is referred to as an ecological island.

The 'island' terminology need not be reserved for an area of land surrounded by water: the island could be a swamp, reef, or meadow surrounded by trees. Mountain tops are islands in that they are separated from other mountains by lowlands. An area of acid soil surrounded by dissimilar rocks forms an island for the species which grow there. A forest, diminished in size by logging, is an island because there is a barrier around it which tends to prevent the spread – both to or from the island – of species. That is the true point of the island concept: it is an area within which a population exists, cut off from easy access by other species.

How do populations relate to confined geographical areas? It can be assumed that a large area can support more organisms than a small one, but what of the number of different species? Firstly, studies show that large areas are capable of sustaining a greater variety of life than small ones. An island may be too small to contain a range of habitats (the canopy level of trees is a different habitat to that at ground level, or below the surface of the soil), or be limited in the rate at which it is colonized; extreme isolation may limit colonization. Animals and plants simply fail to cross the barriers surrounding the island and the organisms found there are therefore less diverse.

This explanation fails to take account of evolution; colonizing species, even if no others arrive, might be expected to adapt to their new environment and evolve into a range of new species. This is precisely the situation which created the different species of finch and tortoise which Darwin found on the Galapagos Islands. Each island was slightly different, and this caused a range of beak shapes to evolve in the finches, based on their food supply. Similarly, fruit fly communities on neighbouring Hawaiian islands have faced evolution in different ways. Of about 1,500 species of fruit flies so far identified in the world, over 500 of them are found only within the Hawaiian chain. Their lineage can be traced to the age of the islands (some of which are divided by lava flows), the older islands possessing more flies than the newer ones in a clear progression of colonization and subsequent evolution.

From research into a variety of species and floral and faunal groups, figures emerge which are surprisingly consistent. Given an area with established, stable populations, there are relationships which can be drawn between the number of species and the area they inhabit. As a generalization, island communities (being small) have higher rates of extinction than those elsewhere, and possess an impoverished variety of species when compared with larger, more integrated areas of habitats.

There are several reasons for this, including how remote the island is (it might be remote to some species, but not others: birds and insects can fly where others fear to tread), and the order in which species have arrived at the island. At its most basic, an insect or predator which arrives will not survive unless its food plant or prey is already established. Even after this point has been passed, new plant or prey species may never be able to colonize the island successfully as they are not given the chance by existing levels of feeding or predation. In terms of island diversity, a rise from immigration will be countered by these 'extinctions'.

In 1883 the island of Krakatoa, near Sumatra, erupted and destroyed its native plants and animals. Almost immediately, recolonization began with seeds carried to the island and by 1933 it possessed 271 species of plants. However, during the same period another 115 species had been recorded but were subsequently lost. A new island near Iceland,

Opposite: Islands, separated from other areas by expanses of water, form ideal areas for the production of new species. This nocturnal Cuban Tree Frog (*Osteopilus septentrionalis*) is a native of Cuba, the Bahamas, the Isle of Pines and the Cayman Islands. Now, man's influence is breaking down these isolating mechanisms: the Cuban Tree Frog has spread, by hitching rides in crates of produce, to Puerto Rico and the Florida Keys. From there, it is slowly spreading through Florida, probably using transplanted shrubs as a vehicle. It is an increasingly common pattern: American Bullfrogs are now resident in Italy, for example, and competition with endemic species may occur. The Cuban Tree Frog is a highly efficient predator, able to devour anything which will fit into its mouth, including other frogs.

National parks, such as at Bryce Canyon in Utah, help preserve wilderness areas but increase environmental pressures due to growing visitation. In the USA over 273 million people per year enter a national park.

Surtsey, arose due to volcanic activity in 1963. Within months of the eruption several species of bacteria, fungi, gulls and plant seeds were present. In two years flowering plants were growing, and mosses within four years. At ten years the species were more diverse (66 mosses alone), but colonization of the island was and is still proceeding, with wide fluctuations from year to year. It will take many decades until there is a stable community present; it will be even longer before the effects of evolution become visible.

What is abundantly clear is that islands offer an important means of producing new species, and the more remote the island the greater the chance that these new species will evolve due to the pressures of natural selection. What is also clear is that there are relationships between the number of species which will be maintained in island communities and the area they are permitted to inhabit. This becomes particularly important when changes to island areas are considered.

Just as a large area will support high numbers of individuals, small areas will maintain fewer. If a large ecological island is reduced to a small one, say by its destruction (encroaching farmland, logging the perimeter of a forest), it is logical that the number of individual organisms will also be reduced. Food plants are destroyed and the pressure on the remaining plants, due to animals which are forced into the area (for whom there is insufficient space), is extreme. Over the course of days, weeks or years after this massive disturbance yet more individuals die. But how are species, rather than individuals, affected?

If an island is reduced to 10 per cent of its original size, the remaining organisms inevitably have a diminished gene pool and are less effective as a species in adapting to change. If they disappear totally it is a loss to the island's diversity; if the island was the only one populated by that species, it is an extinction and a loss to planet earth. Research indicates that an island reduced to 10 per cent of its area (no matter what size it is to start with) will only support half its original number of species. Additionally, once the area of an island drops below 25 square kilometres, over the course of 100 years it will lose 50 per cent of its species diversity.

These figures make for poor reading. Nature reserves and national parks are intended to maintain wildlife, yet are effectively ecological islands. Almost by default, such parks are set up due to external pressures; they exist to avoid the total loss of a rare species due to such factors as encroaching agriculture or industry. As islands, they may not be capable of sustaining their diversity over long periods of time, no matter how good the management. The smaller the area the worse the rate of loss. The Atlantic coastal forests of Brazil now struggle for existence at below 5 per cent of their original area, compared with the time when they were observed by Darwin. It is not enough to feel complacent and mathematically construct a total for the worldwide area of remaining rainforest for, if it is divided into small regions (even if they are designated as nature reserves or national parks), the rate of loss continues in each separate part.

To note that an island is reduced to 10 per cent of its size is a careful, distant, anaesthetized way of saying that individual organisms are dead and half the species are gone, possibly extinct – and, unfortunately, this rate of loss is one that is all too possible in this world. Even if other individuals exist elsewhere in island pockets, the gene pool has decreased. Some alleles will, in all probability, have been unique: there is no guarantee that other populations living in different islands will possess the same mixture of genes; the species survives, but not the genetic variation. Sub-species may disappear; it would be as if the human population of the world (approaching 6 billion) was reduced in number, but lost all the genes carried in its Inuit, Yanomamo and Maasai sub-groups.

Think of it this way. On a day in the future, you arrive in London for Cruft's Dog Show. On entering a vast hall, echoing to animal sounds, you find it filled not with a selection of pedigree dogs, but stall after stall of dachshunds. They're the only dogs left in this imaginary world, carrying the only canine genes that now exist. Alsatians, terriers, pugs, Afghans and what were previously over 500 breeds developed from their ancestral wolf, are gone. Yet, all were dogs and therefore the *species* survives in the form of the dachshund; the dog is not extinct.

The species lives on, but what a loss to the gene pool. Preventing such losses in the real world is as vital to conservation as avoiding species' extinction. Islands provide important opportunities; given genetic variation to work with, here are the centres which produce new species. Removing that variation – reducing the gene pool – does not bode well for the future. Speciation is a slow process that is estimated, at best, to take at least 200 generations. Losing that rich source of diversity may be extremely rapid. Just as extinct species cannot be recovered by nature, neither can genes.

As to the question 'What is a species?', we do not have a perfect answer. In the face of budding and virgin birth, hybrids and polyploids, the concept of an organism which can only breed with its own kind is indeed challenged. Sub-species on the borderline of attaining their own species' status, still capable of interbreeding but restricted by behaviour or ecological islands, muddy the water still further. Perhaps, then, to our working definition of a species should be added a restriction: that interbreeding must take place under natural conditions. Where man has influenced the outcome by deliberately placing dissimilar organisms together, as a test to see what potential for interbreeding exists, that interference rules any issue from those organisms as beyond the biological definition.

In the end, all we have done is refined our flawed concepts, even though those concepts are what biologists rely upon to determine and count the products of biodiversity. Simply, imperfect as it is, the idea of a species breeding only within its own population remains the best definition we have, and with it science must do the best it can.

CHAPTER FIVE

A GENE TOO FAR

'Is it fair to ask whether anyone has given any real *thought to what to do if (and I know it is only an 'if') herbicide resistant crop plants spread to become weeds in fields of other crops? Won't the inevitable answer be newer and stronger herbicides, leading to further reductions in biodiversity?*

'Ladies and gentlemen, I do *think we sometimes need to remind ourselves that our own species has become immeasurably more powerful than any that has ever existed on Earth. We have transformed the face of the land and spread our pollution far and wide, but surely we have, somehow, to find the space and maintain the conditions in which a wide variety of other species can flourish and evolve. I happen to believe that technology should be the servant of mankind and not the master. . . . Yet can we really go on behaving in such a cavalier way and still call ourselves civilised and responsible human beings?'*

HRH Charles Prince of Wales, from a speech given in December 1995

SPECIES, AS WE have seen, change due to environmental pressures acting upon individual variation – a process of natural selection – as well as from sudden alterations in the constitution of DNA. It is equally valid to make a link to the genes which cause those variations, and therefore to the gene pool: the more genes that are available to control each variation, the more variation can exist. In addition, hybrid and polyploid individuals offer other, faster routes to the formation of new species in the natural world, again depending on the genes they carry. Genetic material, then, provides the coded language at the centre of life.

Deliberately forming hybrids by carefully breeding plants and animals can lead to new strains and breeds, in the same manner that hybridization can arise by chance and approach the same end in nature: artificial selection is man's adjunct to natural selection. However, species and strains do not only form under these fundamental laws. A comparatively recent development has been to deliberately and directly interfere with the genetic structure of a wide medley of organisms to produce new varieties or species. A major topic of debate, this procedure is at the forefront of genetic engineering, a discipline where biology meets technology: biotechnology.

The principle of genetic engineering is the modification of a cell's genetic material or the introduction of new, 'foreign' DNA from another organism in a beneficial or useful (as defined by man) manner. A goal might be to cure a disease or prevent its onset, or to fabricate some useful biological substance or product. An example is the manufacture of mice, bred for research, which develop human-type cancers, or modified cells which excrete a cheaper pharmaceutical chemical. The affected cell might be a virus or bacterium, or belong to a plant or animal.

Single-celled organisms with an altered genetic structure will pass their characteristics on to any daughter cells; it is a permanent change which, as there is only one cell involved, must affect all subsequent generations. Similarly, tinkering with the genetic structure of a cell just after fertilization will affect the next generation, as all other cells produced during growth will be identical and based on this embryonic cell. If, on the other hand, cells from part of a developed organism (such as the liver, or a leaf) are affected, no new characteristics are passed on as the reproductive cells are not involved. Any benefits (or otherwise) to the organism will be restricted to that one individual. To cause a genetic change in the offspring the sex cells must be affected. These potential combinations for change in different cell types, at different times in the development cycle of the organism, offer boundless possibilities for man to alter DNA.

This 'future science' has been with us since a California day in 1973. A molecular biologist, Herbert Boyer, was presented with a commercial funding proposal to form a new type of company, Genentech, by Bob Swanson. Today, in California alone, this beginning has grown to over 200 biotechnology companies in a switch from normal research funding, offered by universities and government grants, to that of the private marketplace.

The breakthrough essential for biotechnological advance was the ability to prepare DNA for introduction to another cell. Boyer and a colleague, Stanley Cohen, used the fact that genes are separated from each other by a coded section of the chromosome which indicates a 'start' and 'stop' function for the genetic instructions, similar to the use of a capital letter and full stop to indicate the start and end of a sentence. They determined that certain enzymes, called restriction endonucleases, could be used to cut the DNA into sections at these (or other specific) sites. As the (still attached) stop and start segments were able to join with each other once again, genes could be separated and then rejoined in a different order, turned round before reassembly, or 'foreign' genes could be introduced, like joining together sections of a model railway track in new combinations. The process is termed gene splicing.

The difficult problem of inserting genes without damage to the cell during this 'splice of life' also had to be solved. In normal operation, viruses attack cells by 'injecting' their own genetic material, forcing the cell to perform a dance to the virus's commands. Making use of this inoculation system, so that a modified virus inserts the required DNA, is a valuable technique to the genetic engineer.

A group of viruses called retroviruses are particularly important in this procedure. Retroviruses reproduce in a different manner from other viruses, as they make copies of their own genetic material in a sort of 'backwards' process. Normally, DNA is used as a template for controlling the production of proteins using a 'messenger' called RNA to transfer the data to the cell's manufacturing site. A retrovirus consists of RNA, which produces a DNA copy of itself inside the infected host cell. In turn, this new DNA integrates with and becomes part of the host's chromosome. This retroviral attack sequence forms the basis of particularly insidious diseases because the viral material is carried in the DNA itself and is therefore not recognized by the cell as being foreign; any treatment must therefore take place within the infected cell and involves dealing with a section of chemical which the cell accepts as part of its normal structure. HIV, which leads to AIDS, is an example of a retrovirus.

If a retrovirus is 'loaded' with the desired gene, it can be used to insert that segment of DNA into the host cell, a procedure termed recombination. Whenever the cell reproduces, new cells will be based on what are referred to as recombinant-DNA genes. Electricity and chemicals can also force genes through cell membranes, and specialized 'DNA guns' shoot genes directly into living plants and animals as part of transgenic – transferred gene – techniques. Bacteria themselves may transport the material, for example when *Agrobacterium tumefaciens*, which causes growths called galls, was used to insert selected genes into cotton to improve its yield.

The obvious direction for genetic engineering to move in was the medical field, as this was already established as a market with a proven sales base for expensive products; research had to show tangible results. Pharmaceutical drugs which are based on large, biological molecules can be manufactured, but the process is expensive. Initially, therefore, large protein molecules were targeted by the new gene splicing techniques.

The level of sugar in human blood is controlled by a hormone called insulin (a hormone is a category of chemical which controls other chemical activities in the body). If the production level of insulin is not matched to the actual level of sugar which is present, the blood sugar level fluctuates and becomes either too much or too little, and the patient becomes diabetic. Once the cause and potential treatment of diabetes had been identified, the difficulty became one of finding a viable, artificial source of insulin to give to patients. At first, in 1921, the life-saving hormone was extracted from the pancreas of dogs, and later from pigs and horses, but it nevertheless took until 1965 for insulin to be manufactured in quantity.

A bacterium, *Escherichia coli*, was at the centre of this early success story. The bacterium occurs naturally in the large intestine of all warm-blooded mammals, but causes diseases such as gastro-enteritis if it finds its way to other parts of the intestine. Once the gene responsible for making insulin had been isolated from a human cell, it was extracted from its chromosome using enzymes. A circular piece of DNA (known as a plasmid) was removed from the bacterium, and also split using the same enzyme. The human DNA fragment then recombined with the plasmid to make it whole once again, and the plasmid was replaced in the bacterium. From then on, as it contained the vital piece of DNA, *E. coli* manufactured human insulin and could be grown and harvested as required. This was the first time a human protein had been produced by this technique, and in 1981 insulin finally became cheaply and freely available. The bacterium, *E. coli*, proved so perfect for the purpose that the species is in widespread use as a subject for genetic engineering.

The medical field has continued to benefit from such advances. The interferon drug, used in cancer and some viral treatments (for instance those for the common cold), has been genetically engineered since 1981. Hormones, such as those controlling growth (to treat dwarfism), were produced in 1985. It is the lack of Factor VIII which causes the commonest form of haemophilia; the agent has been genetically engineered since 1988.

Vaccines may also be prepared by modifying plants, in the same manner that cotton yield was increased. In 1995 Texan researchers announced that potatoes and tobacco leaves had been altered to produce human vaccines, which could protect against infectious diseases: eating an engineered potato would, in theory, provide immunity against such diseases as hepatitis B and diarrhoea. One problem is that the vegetable must be fresh and raw, so the team is working towards a genetic banana which could carry a range of vaccines in every bite. The thrust of similar research is towards seeds, as these would not only produce vaccines, but also enable storage without risk of rotting.

The preparation of drugs and vaccines, and screening systems which diagnose diseases such as malaria, are indirect pharmaceutical treatments and comprise about 66 per cent of biotechnological work. Of the other 34 per cent, often unnoticed, biotechnology has already entered our food chains in both real and potential ways; food production and environmental concerns are expanding areas of research that each command some 10 per cent of the market. Sheep have been bred which are larger and have a better meat yield than their predecessors. Fish farms rear larger fish, and pigs, cattle and chickens increasingly follow the same route. Bacteria are grown in volume

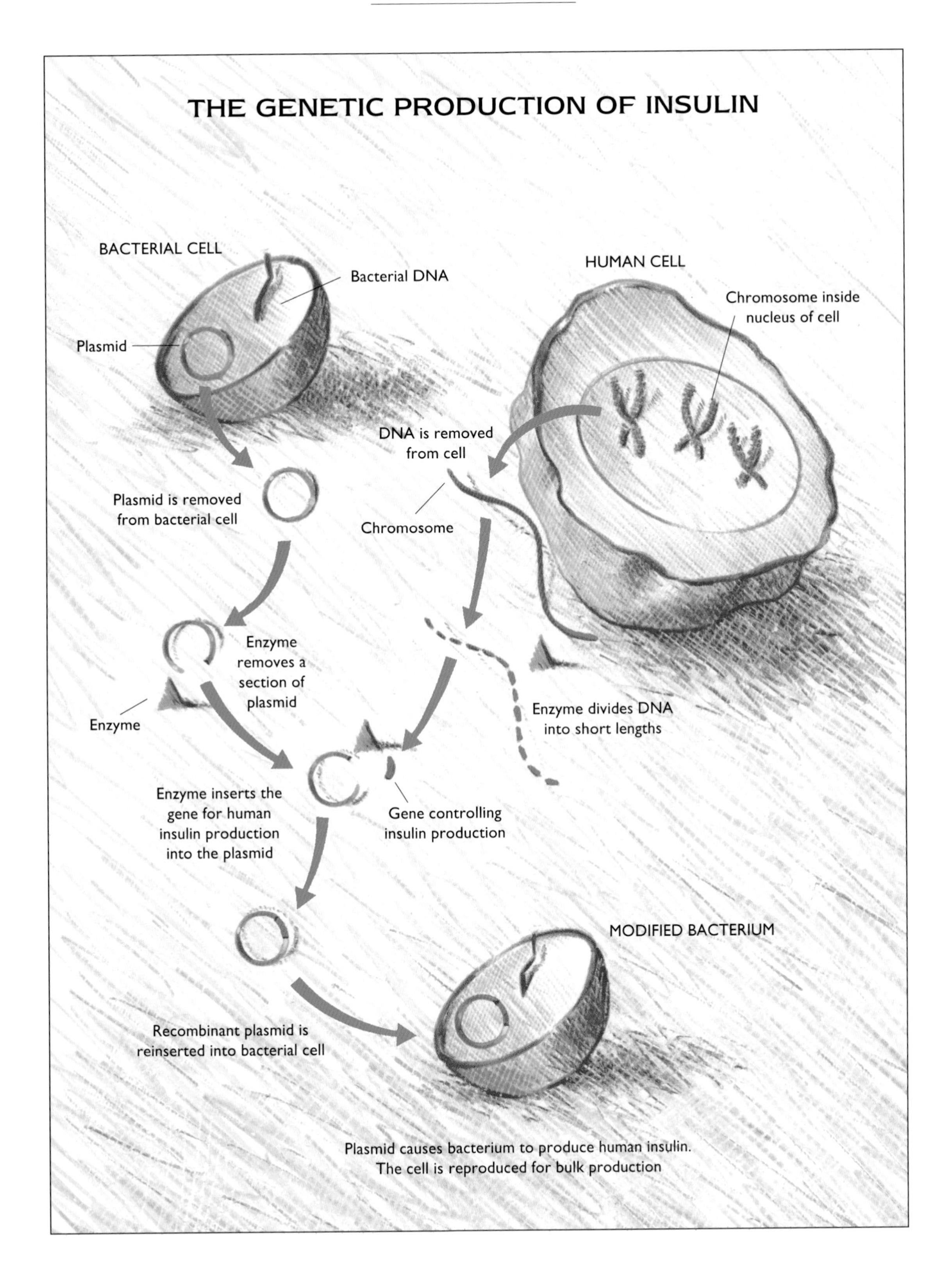
THE GENETIC PRODUCTION OF INSULIN
BACTERIAL CELL
Bacterial DNA
Plasmid
HUMAN CELL
Chromosome inside nucleus of cell
DNA is removed from cell
Plasmid is removed from bacterial cell
Chromosome
Enzyme removes a section of plasmid
Enzyme
Enzyme divides DNA into short lengths
Enzyme inserts the gene for human insulin production into the plasmid
Gene controlling insulin production
MODIFIED BACTERIUM
Recombinant plasmid is reinserted into bacterial cell
Plasmid causes bacterium to produce human insulin. The cell is reproduced for bulk production

on such substances as methane, and can be harvested as food: bacteria consist of over 60 per cent protein, a content greater than soya, beef or fish. Experiments on livestock in the Netherlands and Scotland have yielded genetically modified cattle which secrete pharmaceutical chemicals in milk: the animals are living bioreactors.

Genetically engineered food crops may, for example, benefit from a greater yield and resistance to disease and pests. Some plants can manufacture a poison which prevents pests from eating them, while others (typically, genetically domesticated crop species) cannot. A natural advance, therefore, is to use a gene from one plant species to help protect another. Cabbage plants, for example, require protection against the ravages of the cabbage looper, a caterpillar which eats their leaves. Normal toxins found in plants proved too slow-acting – the leaves were eaten before the caterpillar was killed – so the gene controlling venom production in scorpions was put to work. In trials it proved fast enough: modified cabbages can now be made fit for kings, if not looper caterpillars. A similar experiment conducted in Derbyshire protects apple crops from codling moth larvae.

Some plants, such as clover and other leguminous plants like peas and beans, maintain populations of bacteria in nodules within their roots. These provide the plant with nitrogen extracted from the air, enabling

Oilseed rape has become a widespread crop, providing oils used in cooking, soap, rubber, industrial lubricants, and as a fuel. Rape is only one crop plant which has been genetically manipulated, with some strains no longer able to produce pollen.

clover to prosper in soils containing little natural nitrogen. Again, here is an obvious advantage which some plants possess, but not others. In 1988, after several years of trying, corn was adapted so that it maintained its own nodules of nitrogen-fixing bacteria. The strain is capable of growing in poor quality soil and no longer requires fertilizer, an obvious benefit in a world suffering from food shortages.

The first successful modification to cattle was made in the USA in the late 1980s, when a lactation-increasing hormone was commercially manufactured. The hormone, referred to as BST (bovine somatotropin), controls the production of milk in cattle. Bacteria were prepared which produced BST, which was supplied to cows to increase their naturally occurring level of hormone: milk yield increased by 25 per cent. Bacteria have been modified to manufacture enzymes to digest such substances as oil (a useful ability when tackling spillages) and engineered strains of yeast act on wastes such as wood pulp, at one swoop removing the waste and producing a useful by-product: alcohol. In addition, the bacteria themselves can be harvested and the protein used as animal feed.

In 1995 the British government followed the USA and gave the go-ahead to produce engineered foodstuffs. These include oil from oilseed rape (some strains produce no pollen, a boon for hayfever sufferers), soya bean extract and processed tomatoes. Tomatoes normally go soft as they become over-ripe, the ripening being controlled by a single gene. By removing this section of DNA and reinserting it, reversed, the tomato grows and ripens normally, but then remains firm and suitable for storage.

The essential technique is an old one: by removing part of a system, any observable result indicates the function of the part which was removed. By targeting a gene and removing it, the characteristics it controls in a cell can be determined. This is termed a 'knockout' technique: the gene is knocked out of the chromosomal coding.

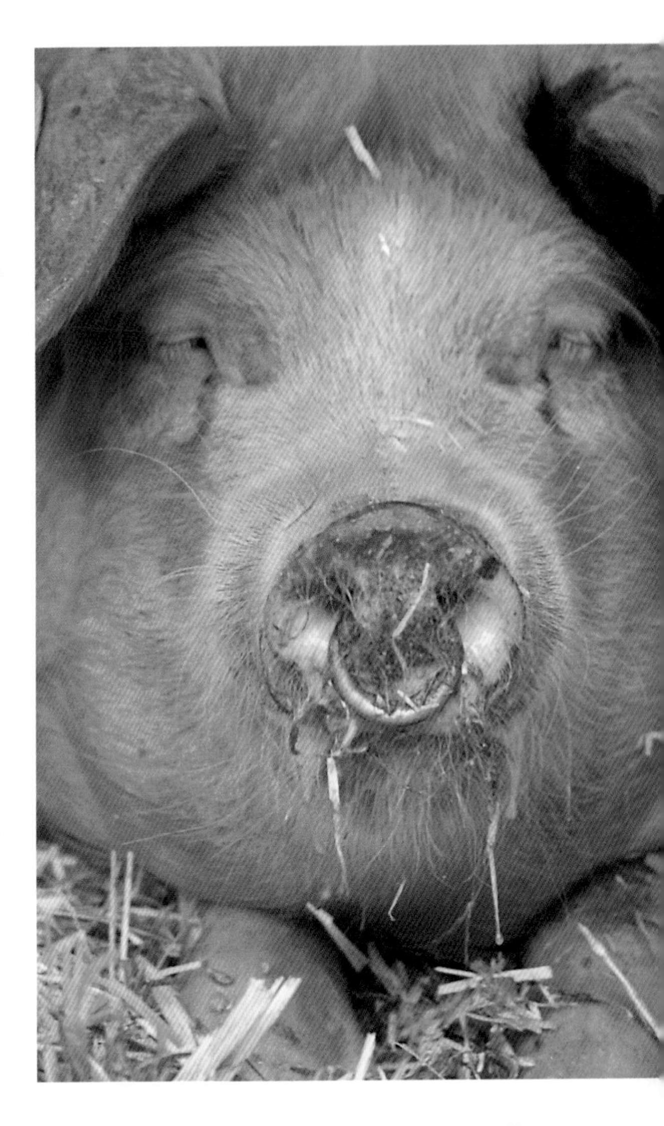

As animal husbandry moves towards genetically modified herds, reaping the benefits of a greater yield of meat in the same manner in which farmers already rely upon specialized hybrid crops, traditional breeds such as this Oxford Sandy and Black pig will disappear.

Knockout techniques are proving useful in research, particularly in the field of evolution. Just as DNA is common throughout life on earth, much of its specific genetic material is similar in different species: there is only a 1 per cent difference between the chromosomes of a chimpanzee and a human, for example. Some sets of genes (called *Hox* genes) seem to be common across a range of animal groups, and control extremely basic areas of an organism's development such as the correct placement of legs or eyes. All mammals contain four sets of *Hox* genes, which seemingly duplicate each other – as if it was easier to copy an existing set than develop new genes from scratch. By creating a duplicate set, extra genes were suddenly available for evolution to operate upon. This may also help to explain sequences of 'jumping' genes, which can reproduce and reinsert themselves into a chromosome. About 10 per cent of human DNA comprises these so-called 'junk' sequences, which may act to reduce the effects of random DNA damage due to invading retroviruses by limiting the probability of 'active' gene sequences being infected.

By knocking out *Hox* genes one by one, it was hoped to discover what they specifically controlled. Some unexpected results arose: mice, flies and every other organism in the programme reverted to an ancient, evolutionary primitive state. During their embryonic development, all vertebrates progress through identical stages in which they possess a tail and neck bulges which approximate to gills: it seems animals store their evolutionary heritage in their genes. When a *Hox* gene was deleted, it caused such throwbacks to appear. Mice avatars exhibited ear bones which were partway between a reptile and a mammal, dating from the age of dinosaurs when the first small warm-blooded animals evolved. When another gene was switched off the mice developed a mammal skull with the characteristics of a fish. The *Hox* code may eventually, via knockout techniques, chart more evolutionary trends than we ever believed possible.

Knockout genes are also revealing how much is under cellular control. As an example, removing a single specific gene from mice results in poor memory. It is feasible to search for genes to remove or add to farm animals to increase their docility and make them easier to control and slaughter. Knockout techniques, in fact, only provide a logical extension to what has already been accomplished by animal husbandry, breeding for specific characteristics, but there are obviously ethical considerations. How far should science go?

While most people will acknowledge the overall importance of genetic engineering, not everyone is in agreement with widespread uses of the technique: even the results of the *Hox* gene experiments are in question (what if the gene deletion is only producing 'faults', instead of throwbacks?). Opponents query whether the genetic control of fruit ripening (to make delivery of fresh fruit to the table cheaper and more viable) or the time of blooming for cut flowers to suit human, seasonal whims, are acceptable uses of biotechnology. In the case of tomatoes modified to control ripening, there is a further finger to point. To check whether the gene has been reinserted correctly, a second gene is added at the same time. This confers resistance to kanamycin, an antibiotic, so that the tomato can be cheaply tested: if it survives exposure to the antibiotic, it must also possess the gene for preventing mushy fruit.

Following the same route, Australian sheep are the subject of experiments to introduce genes which cause the growth of more wool on the animal. At the same time, genes are added which produce chitinase, an enzyme found in cucumbers which attacks insect larvae. According to the theory, if sheep contain this gene the enzymes should destroy any parasites and bring an end to the need for pesticides and sheep dipping. Sheep would remain fit for human consumption; after all, as the researcher, Kevin Ward, noted, 'Chitinase is already in your salad.'

Modified tomatoes are also decreed safe for eating, as the kanamycin-resistant DNA is destroyed by processing. However, a question has been posed by the Genetic Forum: what if the unprocessed tomato *was* eaten? Bacteria reproduce quickly and, like other life-forms, react to the laws of evolution. Would natural selection operate to produce populations of bacteria which are resistant to the antibiotic, just as penicillin is less effective today than when it was first administered in 1941? Farmers, meanwhile, question the use of BST hormones supplied to dairy cows: side-effects in cattle may include decreased fertility and mastitis.

Other, long-term fears surround the ability to grow plants in poor soil or polluted ground. While this has to be applauded, genetic engineering's opponents point out that the incentives for conservation and preventing pollution are diminished. Likewise, as more plants are altered to produce toxins against pest attacks, new genes become available to the gene pool and therefore provide further variations for natural selection. In the future, as hybrids are formed, this could mean the production of new 'super-species' that could seriously threaten biodiversity by eliminating rare species which rely on a limited range of food plants. If further trials with scorpion venom-laden cabbages are successful and the plants become more widespread, any organisms higher up the food chain which are dependent upon the pest species will inevitably suffer.

The production of man-made species could hold more than a dangerous precedent. An early apprehension was the possibility of engineering a species against which man had no defence; it is the subject of countless science fiction plots, based on doomsday diseases loosed upon the globe. If an antibiotic-resistant bacterium or virus was released which caused a lethal epidemic, it could decimate populations with no immediately viable defence. It is technically possible, and the background evidence of natural disease organisms adds credence to the concern. A wave of influenza, sweeping around the world, can infect significant proportions of the human population; through historical times, some 31 pandemics (widespread epidemics) of flu have occurred naturally, that of 1918 was estimated to have led to 20 million deaths. It is not only that a genetically altered microbe could generate these results, but also whether a 'safe' bacterium or virus, inadvertently released, could mutate into a new, virulent form.

Thankfully, such fears have proved ungrounded, even though the first modified bacterium was *E. coli*, itself capable of causing serious disease. During the course of the 1980s, gene splicing became more widespread as restrictions on the technique were relaxed and, by the end of the decade, new species were being released into nature. Food crops were sprayed with modified bacteria, conferring protection against cold; now that the microbes were released, what would be the result on naturally occurring populations of bacteria in the soil? The measured effects were and are slight, but the difficulties can easily be seen.

Fruit ripens, typically changing colour and tasting sweeter (encouraging animals to eat it, and thereby disperse the plant's seeds), under the effects of temperature and the chemical ethylene. Ethylene is released by the fruit in the final stages of ripening, breaking down starch into sugar; citrus fruits remain tart as they contain little starch.

Fruits are normally picked and shipped while unripe, then ripened on demand by gassing with ethylene. In the case of citrus fruits, cold is also needed to produce bright colours. We therefore already control ripening in fruits to suit our demands. The potential in altering genes to further control ripening is vast, though its need is questioned by some people.

Ethically, nowhere is dissension more rife than where human genes are concerned. Just as it is possible to modify bacteria, it is feasible to insert genetic material into human cells in order to prevent or treat diseases. Genes which are already present can be screened, and children's characteristics may be predicted. Screening can therefore provide a person with information concerning the risks of contracting a genetic disease, or the probability of producing offspring which will carry a genetic defect or disease. These may be deformities such as a cleft lip, or serious diseases such as haemophilia, cystic fibrosis or the mentally retarding phenylketonuria. By knowing that children will be affected by a genetic disease, a decision can be made by the parent to begin a family, or terminate a pregnancy – already an emotive issue.

There is a corollary, however: should parents be able to choose the sex of their child – or the eye colour, build, height, skin colour, or any other characteristic from a checklist? Technically, this will be possible as further advances are made, if ethics permit it. In addition, what happens if insurance companies and the workplace use genetic information to decide on what benefits, insurance cover or employment should be offered? A little knowledge can be a dangerous thing.

In 1990 human gene therapy became a reality when a child was diagnosed as having a faulty immune system. Specifically, blood cells could not produce a particular enzyme (ADA: adenosine deaminase) which is crucial for the immune system's operation in protecting against disease; without ADA the bone marrow could not make white blood cells, and any infection (however minor) could cause death.

Some of the child's marrow cells were removed and engineered using a retrovirus, whose own ten-gene code had been altered to render it safe from causing disease. As the retrovirus spread through the cells, it infected each one and caused a functioning ADA gene to be inserted into the marrow cell's own genetic material. These cells were then returned to the patient in order to reproduce and allow the retrovirus to spread; normal ADA production began, and the child slowly improved.

Genetic diseases such as haemophilia, cystic fibrosis and muscular dystrophy may all eventually succumb to this branch of science. Taken to its extreme, the potential exists to design a person and, if the genetic alterations are introduced early enough in the reproductive process (to the first cells which result after fertilization, or even the sperm and ovum), these characteristics will be retained and in turn be passed on to that person's children.

So far, ethics dictate that research has not attempted inheritable genetic modifications. In 1992 a British advisory committee recommended that there was 'insufficient knowledge to evaluate the risks to future generations'. Meanwhile, the parents of children with potentially lethal genetic diseases must await improved techniques and enlightened decisions. Given the opportunity to screen, adjust and modify all material, it is conceivable that diseases, even those such as arthritis, could be eradicated from cultures rich enough to afford the treatment.

This potential alteration in the human gene pool is subject, in essence, to the same arguments applied to the loss of genetic variation in other species. Even genes causing disease may serve a function. Suppose, for a moment, that it was already possible to eliminate the genetic disease of sickle-cell anaemia (research into engineered mice which carry the defective gene, enabling easier study of the disease, is proceeding). Sickle-cell anaemia affects the production of normal red blood cells. As genes occur in pairs (one on each of the paired chromosomes), it is possible to possess either one or two sickle-cell genes. In the former case, mild anaemia results – the condition is known as sickle-cell trait – while a pair of sickle-cell genes produces severe anaemia, and possibly death. However, a single gene also confers resistance to the single-celled *Plasmodium* parasite which causes malaria. In malarial areas such as Africa this is an advantage: the person is better off suffering from breathlessness than

SICKLE-CELL ANAEMIA

Red blood cells contain a chemical called haemoglobin, which enables the cell to transport oxygen around the body. Haemoglobin production is controlled by a pair of genes in each red blood cell, represented here by two letters: HH. One gene is received from each parent. When both H genes are present, as they are in most people, normal blood cells are made and the H allele is retained in the gene pool.

Haemoglobin genes also exist in an abnormal form (represented here by the letter S) which causes the red cells to distort, producing the genetic disease of sickle-cell anaemia. Affected cells lose potassium, collapse into a sickle shape and cannot transport sufficient oxygen: the person becomes anaemic and usually dies young. In this case, all the sickle-cell alleles are lost from the gene pool.

It is also possible to receive a normal gene from one parent and a sickle-cell gene from the other. Some of the offspring's red blood cells form normally, but others become sickle-shaped. The blood cannot transport as much oxygen, but the affected person often does not realize that a sickle-cell gene is present.

The HS sickle-cell condition is commoner in some areas of the world than in others. This is linked to the presence of the parasite that causes malaria, which uses red blood cells in part of its life cycle. Sickle-shaped cells confer some protection against the disease, so have remained in the gene pool rather than being removed during the course of natural selection.

The possession of characteristics due to paired genes is one of the strong points of sexual reproduction, which also enables 'unused' genes to be passed on to the next generation where they may, under the influence of a changing environment, prove beneficial. An example is the Peppered Moth (*Biston betularia*), which occurs in light- and dark-coloured forms. A light colour is more beneficial for the moth when it rests on lichen-covered trees, as it is better camouflaged; dark moths are soon found and eaten. When the Industrial Revolution took place in Britain, causing lichens to die, the dark moths were less easily seen against the bark of trees and the light-coloured moths were eaten. Whether or not a gene confers a useful characteristic depends on the environment, in this case the presence of pollution which kills lichen; with sickle-cell anaemia it is the prevalence of malarial parasites.

dying. Tampering with the genetic pool (animal or plant, human and non-human) may therefore remove useful, as well as harmful, genes from future generations; just what is harmful or useful cannot always be predicted in advance.

Research into the biotechnical production of new strains or species is expensive, but may result in profits of billions of dollars or pounds to the company concerned. It is common practice to patent the results of commercial research, such as a new electronic design, or else competing companies are legally able to use the results without making an investment. In the USA, in 1980, the first patent on a life-form was therefore granted for a 'non-natural, man-made micro-organism'. The bacterium, capable of digesting oil, was the first in a line of over 200 patented species, including cotton plants which resist weed-killers (and therefore continue to grow while surrounding weeds are killed by high-intensity spraying), and tobacco which does not succumb to insect attack.

Then human genes were transplanted into sheep to force the production of pharmaceutical proteins, and a different class of patent applications began to appear. In 1988 Harvard University applied for and was granted a US patent for an OncoMouse, the first patented vertebrate. The mouse is genetically designed to develop human-like cancer within a few weeks of birth, an obvious scientific research tool. The following year the European Patent Office granted its first life patent, a modification to alfalfa, soya and sunflower genes which produces a higher percentage of protein. Three areas of interest were identified as the basis for the office's decision: the application of the research must be beneficial (for example, in reducing disease), the environment should be protected from the effects of the new life-form, and cruelty to animals must be avoided in its manufacture. However, the question remained: should it be possible to patent life itself?

To be permitted a patent, an invention has to exist which is new, useful and 'not obvious to an artisan of ordinary skill' in that area of expertise. Patent applications have been made to cover sections of genetic material, though they are contested: does DNA isolation, identification or use constitute an invention, which is therefore patentable? In particular, can human genes be patented, when they are obviously not in themselves an invention? In 1991 a Californian court ordered that tissue cells removed from a patient during an operation were no longer the patient's property. The bizarre situation arose when a form of leukaemia caused John Moore's spleen to grow to an abnormal size, 6 kg instead of 0.5 kg. After removal, the 'Mo cells' (named after Moore) were found to produce unusually potent lymphocytes (blood cells which combat disease), so these were cultured and subsequently patented.

However, patenting – with the aim of reaping the commercial benefits from research – does not always bring the expected rewards. The OncoMouse proved a flop for Du Pont, the parent company; few were prepared to pay royalties for its use. In Europe, patents granted for herbicide-resistant crops were reversed in 1995, the decision being upheld at appeal. Specifically, the European Patent Office decided that the patent convention did not permit royalties to be earned on life-forms, regardless of whether or not they were conventionally bred or genetically modified. In Europe, it is no longer possible to patent life (unless the law changes once more). Neither is gene therapy always successful; immune systems present problems in rejecting altered cells, and modifying the correct cells may sometimes prove difficult. Of the trials conducted on human volunteers, complete cures are rare. In medicine, genetic engineering does not yet offer a panacea.

In the late 1980s, as DNA investigation techniques grew cheaper, an immense project to map the position of all human genes was begun. Genes may differ between individuals (blue or brown eyes, tall or short), but the genes for any one characteristic always occur at the same place on the chromosome – like railway stations on a map which never move in relationship to each other. Named the Human

Genome Project (the term 'genome' refers to the total genetic composition of an individual organism, or of the gene pool of one species), it is a 15-year plan as not only must the genes be identified, so must the sequence of nucleotide bases, an estimated 3 billion of them, which form their coded language.

However, no two people are identical – that is, their genes differ – and while the basic genetic map will remain the same, the alleles at each site may be dissimilar: *this* person has the ability to detect a certain bitter taste, while *that* person cannot. There are two distinct alleles involved; the genes controlling many characteristics are present in two forms, such as the ability to roll the tongue into a tube or not, but other gene sets are more numerous. Three alleles control human blood groups, although any one individual may possess only two of these genes (one on each chromosome of the pair). The situation is multiplied to astronomical levels by each of the possible 100,000 gene positions, any of which may be filled by an unknown number of genetic variants. To use the earlier analogy of picking pairs of marbles from a bag, in this instance we must correctly place 100,000 marbles in a precise sequence on each chromosome. Worse, for each of the 100,000 positions there is a set of different marbles to choose from – and the number in each set is unknown. Only by repeatedly dipping into the marble bag can the marble types and their relative numbers be determined.

There are therefore more genes available in the population than can be carried by one person and, given this, the Human Genome Project could never hope to identify all genetic variation. A separate organization was therefore proposed in 1991: the Human Genome Diversity Project. This is attempting to map the genomes of the world's ethnic groups (which are expected to contain genes for uncommon traits), some of which are fast disappearing. With them goes their genes, and thus the human gene pool itself is diminished. Top targets are groups such as the Hadza of Tanzania and Yukagir of Siberia, the potential rewards being answers to questions of humanity's ancestry and spread across the globe, and the possibility of 'unusual' combinations of genes for medical research.

Genetic variation is not restricted to the present; extinct species may yet yield their genomes to science. In the 1993 film *Jurassic Park*, *Tyrannosaurus rex* struts across its compound, and resurrected creatures fly and run around their island amusement park. The film's premise revolves around reconstructing complete chains of DNA from sources such as insects trapped in amber, and substituting modern DNA from amphibians and birds (the descendants of dinosaurs) where sections are missing; modern DNA differs little from its ancient predecessors. It was science fiction based on fact.

In 1984 American scientists Allan Wilson and Russell Higuchi isolated DNA from the

Birds, such as this Snowy Egret (*Egretta thula*), are sometimes referred to as 'descendants of the dinosaurs'. Although their characteristics of warm blood and feathers separate them from reptiles, birds still possess tell-tale scales on their legs which link them to their ancestors.

dried muscles of a preserved quagga, a zebra-like animal which became extinct in 1883. The technique proved highly interesting, as it enabled direct comparisons of DNA with those of modern animals, and helped answer questions of their ancestry. Other experiments soon followed, in which short strands of DNA were removed from a 2,400-year-old Egyptian mummy. In 1988 a giant ground sloth found in a Chilean cave yielded the oldest DNA yet recovered at that time: 13,000 years old. Subsequent work showed that some fossils can hold preserved DNA for much longer than previously thought, and in 1990 genetic material was extracted from a fossil magnolia leaf. At 16 million years old, this opened up immense possibilities.

There are problems with such preserved material. Early attempts were foiled by bacterial DNA contamination and, even after taking several samples, the fossil DNA could not be reassembled to form anything useful. Then, using a fresh approach, the fragments were unravelled from their helical shape, and enzymes remanufactured links to form new chains. If amber-preserved cells and life locked within stone can yield the information, perhaps *Jurassic Park*'s fiction will not remain that way forever. In 1995 a claim was made that bacterial spores from 40-million-year-old bees caught in amber had been reanimated. If the possibility of resurrecting extinct life seems far-fetched, it might be preferable to manufacturing a genetically tailored new species. Or, is this simply a new, cutting-edge technique which will enable the gene splicing scientists of the future to gain access to banks of extinct DNA taken from today's forests?

Jurassic Park's plot called for dinosaurs, all of which were believed to be female, to unexpectedly breed and increase their numbers after some individuals had switched their sex to male. A number of species can change sex during development, such as clownfish, parrot-fish and wrasse. Some oysters alternate sexes, switching between male and female, while a large number of species are true hermaphrodites (earthworms are one example): they bear the organs of both sexes. In yet another reproductive technique, some species can produce offspring from a single female parent in what is termed parthenogenesis; being genetically identical to their parent, the offspring are termed clones.

Parthenogenetic development is common among bees and ants. In these cases, the queen lays unfertilized eggs which develop into drones. Some species of weevil (many are major pests) can produce young from unfertilized eggs and, in insects such as thrips and aphids, males are extremely scarce or totally unknown.

Parthenogenesis in vertebrates is rare. *Lacerta saxicola* and *Cnemidophorus uniparens* – race runner lizards – are entirely parthenogenetic, while the unfertilized eggs of some frogs (*Rana japonica*, *R. nigromaculata* and *R. pipiens*) can produce adults after being pricked with a needle, as can chickens and turkeys. Genetic selection has been used to increase the rate of parthenogenesis in some turkeys in order to produce fertile adults from unfertilized eggs. Mammals such as mice and rabbits can also be initiated by these techniques, but as yet experiments have failed to bring young produced from a single parent to birth. Parthenogenesis may have future applications with regard to cloning rare species, if science takes that course of research.

Biotechnology holds wonders for the future, and has given us the best evidence we have that life comes from a common ancestor. In one experiment, animal genes were used to modify a species of plant. A single gene controls the production of an enzyme which causes light production in fireflies; when it was inserted in tobacco DNA, the plants became capable of glowing if the correct chemicals for the reaction were provided in the soil. If

Opposite: Clownfish form close relationships with sea anemones, living within their tentacles. Until recently it was not known why the anemone did not attack the clownfish. It was then found that the fish are coated with a sugary layer which picks up anemone cells: the anemone cannot distinguish the fish from parts of itself. Clownfish begin their lives as males, the dominant fish becoming a female as they pair off for breeding.

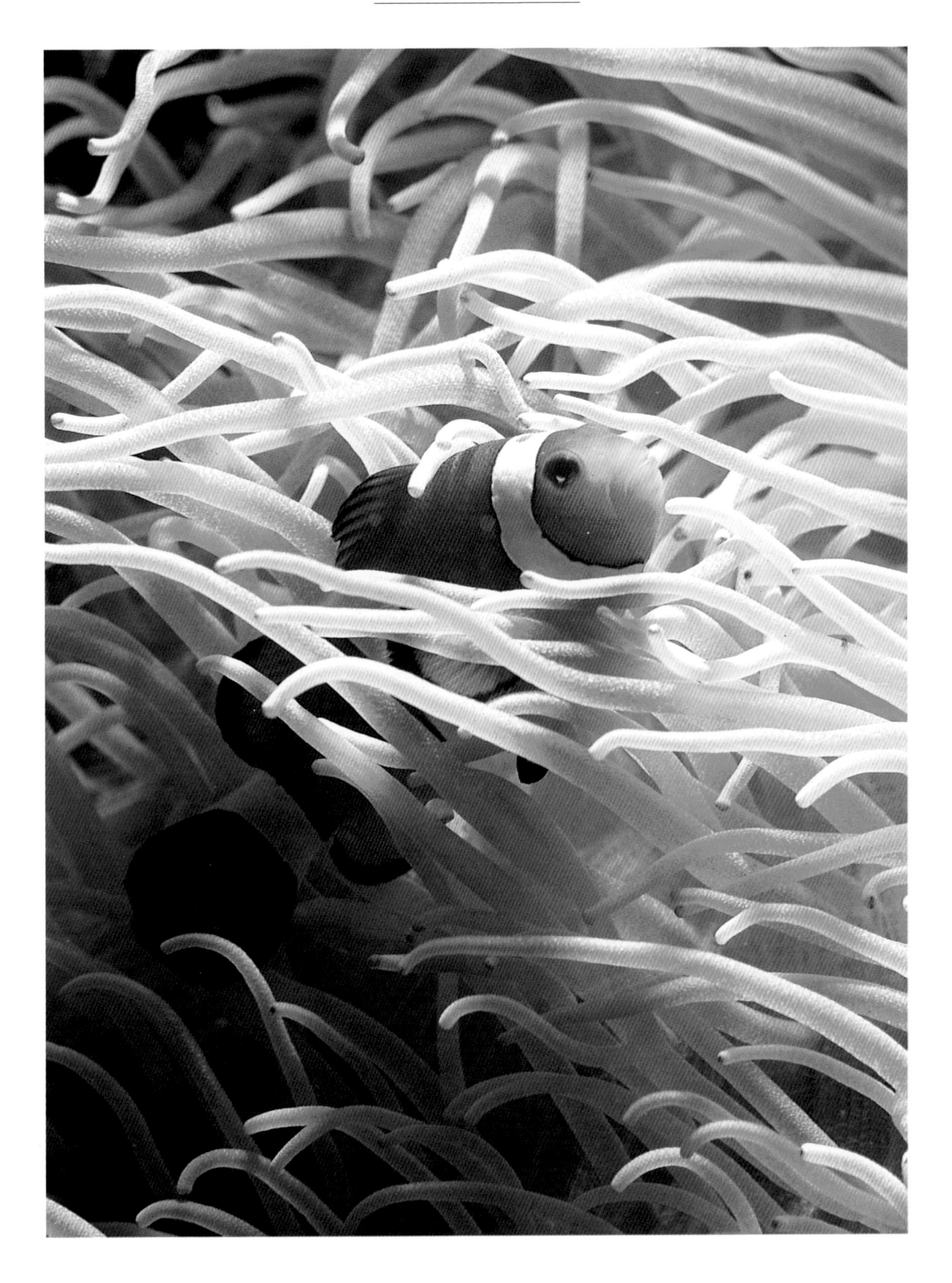

genetic material can be switched from one organism to another, even from animals to plants and vice versa, this is a surest sign we have that life began its evolutionary path from the basis of a single molecule.

Where there is the impetus of commercial profit, science fact exceeds that of fiction. Biotechnology has given us the ability to sample DNA for forensic fingerprinting, even allowing matches to be made between fragments of the Dead Sea scrolls. The scrolls, based on organic material, were preserved in caves since the time of Christ. However, the manuscripts are in thousands of pieces, but are being reassembled using their DNA content as keys to which fragment belongs with each scroll. Culture techniques now grow human skin and cartilage; by using the patient's own cells, there is less risk of rejection of tissue. In 1995 it was announced that an ear, to be grafted onto a child, was being grown in a laboratory, and effective heart valves and blood vessels are on their way.

For transgenic organ production (rather than single tissues), pigs offer more than bacon. In Britain, pigs – chosen as their organs are human-sized – have been bred using implanted human genes to provide a source of donor organs. Although the organs are based on human genes and there should therefore be little problem of rejection, pig-grown organs form attack points for immune systems because pig genes are also involved. For this reason, during tests some of the pig's immune system is also transplanted. It may only be time before the wealthy can maintain a source of organs based on their personal genetic material, to face the day when an organ replacement might be required. Perhaps this is the ultimate way of making a pig of yourself, though there is one obvious danger: with inter-species transplants comes the added risk of encouraging the possibility of inter-species diseases.

Biotechnology is also moving into the realm of engineering and computing. DNA, reduced to its essential function, forms a series of coded instructions. Similarly, computers operate on the basis of bits of information: an 'on/off' code. Leonard Adleman at the University of California has shown that genes can be used to provide solutions to mathematical problems, based on the order in which they are assembled. A molecular readout of the results demonstrated the inherent possibilities of the process in storing computer programs. Organic molecules offer the ability of storing more information in a smaller space than today's silicone systems; from the University of Illinois comes a molecule based on six carbon rings joined in a circle, with a central carbon ring which can spin to one of two positions based on electronic or magnetic impulses: an on/off switch. Bacterial proteins form sophisticated image detectors, and yeast can be used to produce miniature semiconductors.

Biotechnology is a science which has the capability of travelling far along the road of life. All these accomplishments, and more, lie with the use of genes. As more genetic material and the characteristics under its control are identified in plants and animals, the anticipated success rate in finding solutions to some of the planet's ills rises. It may even prove possible to manufacture genes to order; operational artificial genes were constructed as far back as 1973, the technique indicating the possibility of altering genes within a cell rather than inserting new material. As gene splicing and the industry it has spawned move towards adulthood, it is difficult to identify areas of research which cannot be tackled.

However, it remains to be seen if the planet's surviving biodiversity is sufficient to supply genetic engineering's demand for raw materials: an increasing variety of genes. What is certain is that, if those essential genes are to be readily available to future science, conservation of the world's biodiversity is imperative. The industry is powerful, and already approaches a turnover of $40 billion in saleable medical products alone. It is time for some of that profit to be meaningfully returned to the soil which spawned the vital genes which it relies upon.

The time of Genesis is past; man now plays God in the matter of Creation.

CHAPTER SIX

LIFE BURNING BRIGHTER

No man is an island entire of itself.

John Donne, *Devotions Upon Emergent Occasions*, 1624

IF ANYTHING IS guaranteed to fascinate a person involved with the natural world, it is the sheer volume of variety which is found within it. Yet, while it appears so different, from the simplest of lowly worms to the gigantic redwoods of the northern Californian coast, all life bears factors in common. Cells use identical enzymes, and are based on identical DNA. Yet, over the millennia, life has yielded to environmental pressures to form today's cornucopia of form and function. In each case, mutations have introduced new characteristics ranging from chemical structures to camouflage and warning colours, while natural selection has weeded out the beneficial from the irrelevant and harmful. In the words of Cotton Mather (1663–1728), Witch Hunter of Salem, 'What is not useful, is vicious.'

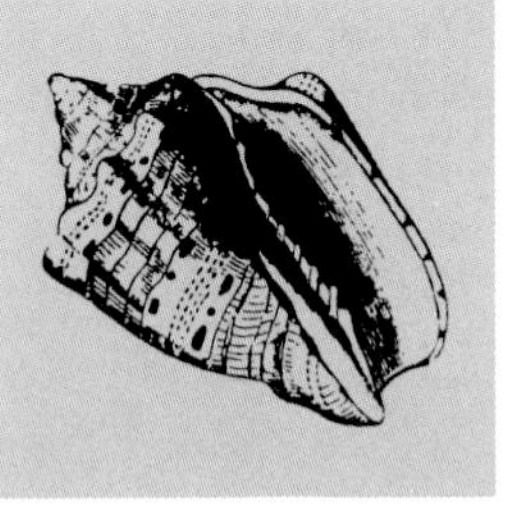

As Darwin's theory of evolution has shown, these adaptations are inexorably driven by the environment – and the greater the stress, the faster the change. More specifically, an organism is moulded by the habitat it finds itself in, be it tundra or veldt, a pond or desert. What is suited to one is, by default, less suited to another. When a new characteristic, or extreme of a range of variations, is randomly tested by the surrounding forces of nature, it must reveal an inheritable advantage or be removed from the chain of successive generations. Evolution – the mode of formation of new life-forms – is therefore driven by the habitat, which includes not only physical factors such as rocks, soil and climate, but also the living things with which the organism must interact and compete. Paul Simon's claim-in-song, 'I am a rock, I am an island', removed from humanity, is ultimately as false for biology as it is for the needs of man. Simon's inspiration, the metaphysical poet John Donne's reflection that 'No man is an island entire of itself', is somewhat more accurate.

All species occupy a specific niche, the place of optimum suitability for that form of life. The niche might be constrained, such as an insect passing the whole of its adult life within the underside of a leaf where levels of protection, temperature, camouflage and water supply are the 'best' that can be obtained, or broad such as a lion roaming African plains, yet staying within a limited set of environmental conditions. When other living things are taken fully into account and form a community of web-like chains of what relies on or feeds upon what, the situation becomes far more complex.

To take a simplified example, consider the effects of sunlight. The sun's energy irradiates the earth and is absorbed by green plants to energize the production of sugar in leaves. This process of photosynthesis has not only manufactured food for the plant, but also for the herbivores which eat it – albeit to the plant's detriment! In turn, the plant must find ways of protecting itself, with thorns or poisons, while making use of animal life in helping disperse pollen or seeds. The herbivore, meanwhile, is trying to avoid being eaten by carnivores, and greater sensitivity to this threat (wider vision, better hearing, to say nothing of a fleeter foot) becomes an evolutionary survival factor. Then, the carnivore must protect itself against other carnivores, yet be cunning and able enough to capture its own food, and . . .

The theme is endless, though this food chain of events – producer (plant) to herbivore to carnivore – is not. At each stage the animals are gathering food for energy, and at each stage there is a loss of available energy so that food chains cannot continue ad infinitum. A shrub eaten by an insect, in turn eaten by a small bird which succumbs to a bird of prey, forms a food chain which cannot continue. To extend the chain beyond the bird of prey is not possible, for a greater hunting area is then required to supply sufficient energy to the last in the line, and this area must have logical limitations.

Whatever form these limitations take, few of these nutritional sequences – these food chains – extend beyond about five organisms, or else the energy flow from plant to animal to animal

Just as animals use plants as a source of food, plants use animals to aid their reproductive process. This fly has visited a rhododendron flower for a free lunch, but will pay the price as an unwitting messenger, spreading pollen to other flowers.

becomes too weak. At each step there are chemical reactions involved and, for all its resilience, life is not particularly efficient – at least, no more so than other heat engines such as automobiles that burn fuel. At best, only about 10 per cent of the energy contained in food is passed on to become locked into the new organism in the form of new muscles or leaf cells: the rest is lost, mainly as heat. Just as a car engine must have its waste heat removed by cooling systems, so must living organisms – if the mammalian liver did not have the heat from chemical reactions removed from it by the bloodstream, it would quickly, internally, cook.

So, space, water, finding a mate: all form competitive elements but, as the existence of food can be traced back to sunlight energy, life on earth ultimately depends on the sun. We humans cannot grow fat while soaking up the sun's rays in the hope of developing a suntan, but we as surely use its energy when we grow, run and talk. In areas of the world where food supplies are at a premium, it is plants which are eaten rather than animals: it is a more efficient use of the nutritional energy flow, with less lost to the system.

Such factors fall into the province of ecological studies. Although this feels like a modern concept, the term 'ecology' was coined as long ago as 1869 by a German biologist, Ernst Haeckel, and elements of the discipline had been under discussion for years before that. Certainly, it has long been recognized that the net effect of these biological interrelationships is that communities are necessarily constrained in a variety of ways, including the degree to which individuals are interdependent on one another.

This vision of a community existing together is a difficult one to study in depth. To do so may involve so many factors that, although elements of it can be understood to some degree, others may prove elusive. As one question is answered, two more take its place; as one relationship is revealed, others are hinted at. The concept of an intertwined web of life is a real one, and it is therefore difficult to predict exactly what will happen if any one element within it is altered. For this reason, closed ecosystems (ecological systems; interdependent organisms segregated in some way from interfering elements) are sometimes studied in the hope of reducing the number of competing elements: to play and predict the outcome of a game of noughts-and-crosses is simpler to achieve with certainty than one of chess. In addition, closed ecosystems sometimes teach us unexpected things.

This story begins with Nicolae Ceaucescu, President of Romania. In 1986, at his seemingly arbitrary order, work began on a new power plant situated on a limestone plateau at the edge of the Black Sea. A shaft was dug which, at a depth of 18 m, intersected a natural cavity. Cristian Lascu, a caver and geologist, made the first exploration and found a short cave which led to a lake of sulphurous water. The air in the cave was laden with carbon dioxide and rank with the stench of sulphur dioxide (better known to schoolchildren for its use in stinkbombs), and subsequently forced the use of breathing apparatus on its explorers. What was entirely unexpected, however, was the rich community of life that Lascu exposed in what was named Movile Cave.

Life is not confined to the sunlit surface of planet earth; deep ocean trenches at nearly 11 km below the waves are capable of sustaining life, and bacteria live in sunless ecosystems hundreds of metres below ground. As

Over thirty species found in Movile Cave have proved new to science, providing a basis for Romania's issue of this set of six stamps in 1993. The spider *Lascona cristiani* was named after the caver who first explored the cave.

to caves, most possess highly developed organisms which are well suited to their stygian environment. The unusual factor about Movile Cave is its isolation from the input of external energy. Sunlight is not involved in Movile's food chains. Thermal vents, emitting heat and sulphur compounds into the sea, have been studied and are known to support microbes that use sulphur as a source of energy for respiration. However, in the ocean there is a ready supply of changing water and nutrients, and in normal cave systems life is supported by an input of energy in the form of detritus washed in from the surface. In Movile, ecology has taken a step further for here, in the dark, bacteria obtain their nutrients from sulphur, with no surface interaction. Even the water does not appear to have a ready connection to the outside, as indicated by a search for contamination six years after Chernobyl's nuclear reactor was destroyed: the results were negative.

This form of nutrition, based on the breakdown of a chemical, is termed chemosynthesis, as opposed to the light-driven process of photosynthesis. It is a less efficient method, relying on extracting the chemical energy locked in sulphur bonds (hence the high atmospheric sulphur dioxide content of Movile Cave). Such systems are involved in bacterial breakdown of basalt deep beneath the earth, as has been discovered in Washington State, where energy is derived from hydrogen produced when iron-rich silicates break down in the presence of water. To begin such food chains, no organic matter is required.

Lascu and a colleague, Serban Sarbu, discovered that bacteria, floating on the surface of the lake as a red or yellow scum, were oxidizing sulphur to produce glucose. Other organisms – nematode worms, small crustaceans, snails – fed on the mats from below, while fungi grew up to be grazed by collembola (primitive, wingless insects which often inhabit caves). Their huge numbers, over 1,500 individuals per square metre, were able to support predatory spiders, centipedes and pseudoscorpions.

In addition, over thirty of the species found in the cave (some 75 per cent of those so far discovered) were labelled as new to science. They evolved without the impetus of changing forces, losing eyes and pigments to not only survive but thrive in an atmosphere which would kill their terrestrial cousins. This closed world may ultimately provide answers to some evolutionary questions, as DNA analysis indicates that the cave's ecology has been sealed to adapt in its own manner for between 1 million and 5 million years.

Even though Movile's food chains of potential prey and predator, energy producer and user, are restricted because of the number of species involved, this does not imply that it is a simple relationship and even here, after much study, there is a great deal to be learned. Imagine, then, the difficulties in translating the study of ecology to a 'free' world where organisms can come and go, traverse mini-habitats and ignore or interact with perhaps thousands of other species in uncounted numbers of individuals.

The complex way in which organisms interact with each other in their feeding chains builds up to what is termed a food web, a convenient way of displaying the information which has been gleaned, to aid understanding. In a food web all feeding relationships are shown, and inferences begin to be gathered.

This extremely simple example of a food web (opposite) offers a number of possible food chains: sunlight energy supports the vegetation, which is eaten by the snail, which gives its life and energy to the thrush and hence to the kestrel. A different chain runs from the vegetation to the rabbit and then the fox. The direction of the arrow is specific: it shows the energy flow.

This web of life is indeed simplified, for there are many more possible sources of food for the organisms involved. However, it does serve to illustrate the consequences of change. Suppose that a disease decimates the rabbit population. It is readily apparent that the foxes and kestrels are suddenly restricted in their diet and come into stronger competition with each other for food resources; they will each

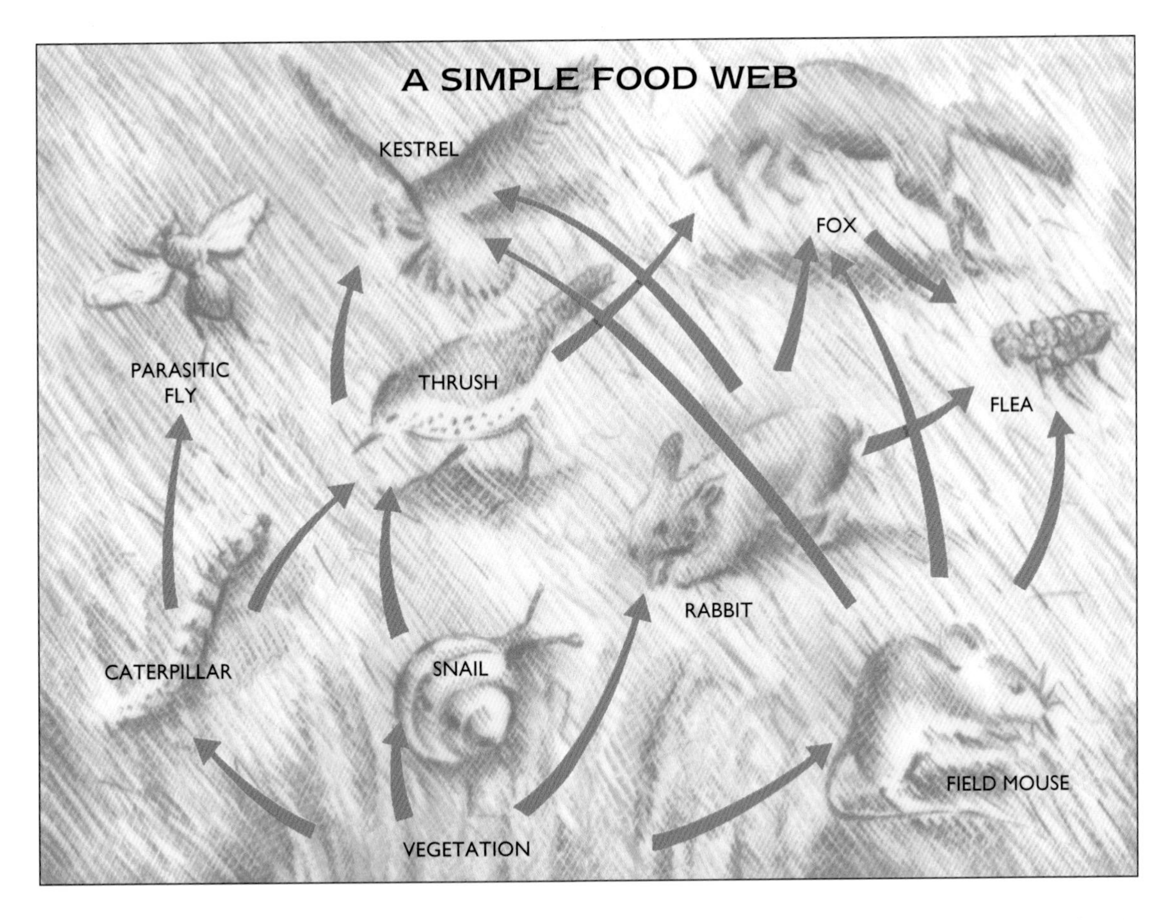

attempt to consume more mice and birds. However, as these animals decrease in numbers and food becomes still scarcer, the foxes and kestrels must diminish in numbers – but, at the same time, with less predation, the caterpillars and snails will increase in numbers. Ultimately, in this hypothetical situation, insects such as the parasitic ichneumon fly will benefit and increase its population. It is a perfect example of how a single change in a natural system can have far-reaching consequences which, all too often, are virtually impossible to predict in the real world.

The balance of nature is fine; tilt the support, add an off-centre weight, and the fulcrum-balanced load comes tumbling down. The organisms which live together in a community have done so for millennia, adapting to each other and successive change, moulding to their niche. Change an influencing factor (virtually anything in the habitat) and the knife edge balance becomes more precarious.

In one experiment a Pacific coast seashore food web was examined. Starfish were found to thrive on molluscs and other sessile organisms on the rocks, such as barnacles. The starfish is what is termed a keystone organism; its effects within the community are vital to that community's balance. As might be expected, when all the starfish were removed their prey – the mussels, limpets and barnacles – increased in numbers. However, these extra shell-encased animals were now in direct competition with each other for space and, with the removal of the starfish predators, some proved more adept in the competition than others. They increased in numbers out of proportion to their neighbours and the original fifteen species in the area were soon reduced to eight, the race winners benefiting at the expense of the losers. The starfish, the keystone species, ensured a greater biodiversity than if that predation ceased.

Any community will develop towards a climax if nothing alters its inexorable progress.

These starfish, on the Pacific coast of Washington State, are keystone species in the seashore food web. Each is a predator, feeding upon shellfish and, in so doing, preventing any single species from dominating the others. If keystone species are removed, either deliberately or as a consequence of man's actions, the unchecked competition between the remaining organisms inevitably results in a decreased biodiversity.

The fate of British land is to grow oak forests, if other factors do not intrude. Grazing and predation are two possible factors which prevent this eventual domination: sheep and rabbits may retain meadows by grazing the heads from oak seedlings, just as the starfish kept the seashore fauna in check.

Predation as a force is therefore vital to biodiversity: without its challenge, the shifting number of species decreases and, in the long term, the food web suffers. In our hypothetical food web, the permanent removal of the rabbits leads to an increase in parasitic flies, but of course this will continue into successive years when the added burden of predation reduces the number of caterpillars. Predictably, the numbers of flies must then decrease with the loss of their prey, which has conceivably been decimated. Fewer or no caterpillars implies fewer moths and butterflies, which some species of plant may depend upon for pollination and, eventually, the diversity of vegetation itself might be affected.

Of course, this is an oversimplification and it seems that nature has many fail-safe mechanisms. In reality, the area in question is not totally isolated and it is unlikely that the rabbit population would be totally destroyed in the first place; fresh stock might invade from elsewhere, and the system comes back into balance. In addition, the web is far more complex than this and other food sources and pollinating mechanisms will exist. However, if the system is stressed too far the outcome will indeed be disastrous. Today's Africa may be low on plant species due to the effects of severe drought in the geological past, while South-East Asian typhoons apparently restrict diversity: similar forests elsewhere in the world, not subjected to this limiting factor, contain more species. The greatest change derives from one source: man is adept at meddling with food webs in a desire to mould nature to what is perceived as desirable. In so doing, he has frequently failed and, even worse, too often not learned from the outcome.

The effects of change can be insidious. In 1907, deer living on Arizona's Kaibab plateau were preyed upon by coyotes and wolves but, in the interests of permitting the 'preferred' deer to proliferate, the carnivores were deliberately removed by trapping and shooting. The deer population indeed rose, from 4,000 to over 100,000, in the course of which the vegetation was stripped and, when the food eventually ran out, the population crashed to a lower level than it was at originally.

When the Australian continent was settled, sheep were introduced and soon became an important source of meat and wool. They prospered in the rich grasslands of the east. Around 1860 seven rabbits were released, ostensibly for hunting. The rabbits also prospered, but competed with the sheep for grazing; over the coming decades they became so numerous that sheep ranches, and therefore the yield of wool, suffered. There was little to keep the rabbits in check, for there was little natural predation other than the Dingo, Australia's wild dog, which itself is thought to be an aboriginal introduction, and the Red Fox, another settler-introduced animal.

In 1887 some 15 million rabbits were killed by settlers in New South Wales alone, and a prize was offered for a solution to the menace. Louis Pasteur suggested releasing a chicken cholera virus, but this was rejected as being too dangerous. Then, in 1897, the rabbit disease of myxomatosis was discovered in South America. The myxoma virus that causes the disease affects the rabbit's mucus membranes; the rabbits become blind, deaf, and there is a 99.8 per cent mortality rate.

Myxomatosis is transmitted to rabbits by a vector, in the same way that malaria is transmitted to humans by mosquitoes, and in the fourteenth century the Black Death (bubonic plague) was spread by fleas through Europe, killing as much as half the human population. The myxomatosis vector was identified as the mosquito, which offered a seemingly perfect opportunity for Australian scientists. The virus was released into the rabbit population in 1951, and proved particularly successful in moist swamps and river regions as this was where mosquitoes bred. However, as these wet areas are limited in Australia, the myxomatosis solution to the rabbit overpopulation was not as widespread as intended.

Hearing of the project, the Frenchman Armand Delille conducted similar experiments, but the virus escaped and spread through France to jump the English Channel in 1953. In Europe, the flea proved to be the myxoma vector, and the rabbit population suffered immense, almost unrestricted losses. Unlike the situation in Australia, where habitats controlled the vector, in the close confines of rabbit burrows fleas were easily passed between individuals.

With the loss of rabbits as prey, established food webs collapsed but, again, the results were unexpected. Stoat populations, which hunted rabbits, diminished rapidly. With fewer rabbits grazing the seedlings, shrubs could grow to maturity and grasslands became luxurious. In turn, voles increased because of the extra concealment and food offered by the vibrant growth, but, as voles are hunted by weasels, these increased in numbers. By the time some semblance of balance had re-established itself, decades later, the shrubs and trees were beyond the point where they could be materially affected by rabbit grazing (or by their man-controlled replacement, sheep), and new communities had become established. In the new woodlands mice and voles prospered, supporting more foxes, which raided more farmers' poultry. Cause and effect, the confused results might only become observable years down the line.

In Australia, history seems about to repeat itself as, in 1995, an experiment begun on Wardang Island permitted a rabbit calicivirus to escape to the mainland. It began to spread at up to 8 km per day in an unstoppable surge, decimating populations of rabbits. Rabbits are the main source of food for foxes, which can be expected to turn to indigenous species when the rabbit population falls; in addition, Australia is now the world's largest exporter of fox fur, an industry which will suffer when, in turn, fox numbers drop.

Physical alterations in the environment can also introduce new forces, sometimes by promoting growth. Take, for example, the addition of nitrates to the soil to increase the yield of crops, or the discharge of clean warm water from industrial processes. When nitrates or heat enter lakes and rivers, water plants also benefit and make enormous spurts of growth. Unfortunately, other nutritional requirements may not be present, and grazing does not control this rapid growth so that, when the plants die, there is a massive pollutant effect. As the plants decompose, oxygen is used up and the other water-borne life is suffocated.

Likewise, where tropical irrigation canals are established, transporting life-supporting water to crops, perfect conditions are established for snails and other invertebrates. These expand in numbers, occupying new territories and colonizing new ground. However, while some communities benefit from the increased water supply (and man from more crops), snails can transmit the disease of schistosomiasis. A direct link between increasing irrigation and a rising incidence of the disease, currently affecting 250 million people, has been found in the Third World. Elsewhere, water hyacinths have entered the ecosystem and grow rampant in irrigation ditches, slowing water flow and clogging navigable canals. Introduced to the USA from South America, in areas they have overwhelmed native species and are called the 'beautiful nuisance'.

Predicting the exact outcome of changes within habitats is an awesome task which man is far from capable of achieving with any certainty; the most that can be hoped for is a probable effect. What is certain is that, to be able to make any form of prediction at all, the totality of an ecosystem has to be studied. Take an example of a lake. It is not enough to know only the physical factors, such as temperature at different depths and seasons, or what nutrients or pollutants are present; the community's interactions also have to be known. What will happen if a new organism is added to the community, or removed from it? Will it survive, or not? Will its occupation of an empty niche where it is not in direct competition with its cousins result in greater diversity, or will there be a ramification which leads to a diversity decrease?

As has been seen, a prediction with any fine degree of accuracy is still impossible to achieve, but clues can be gleaned by studying some of the forces which pushed today's life through its evolutionary course. What we have discussed so far has been the result of interacting species in their habitat; a perhaps greater, if coarser, effect was the separation of populations in the geological past. By default, organisms must exist within a habitat of one form or another; what was the major influence which created these different sets of conditions and set the parameters for evolutionary change?

The segregation of life due to Movile Cave's 5-million-year-old entombment is a mere pinprick in geological time, when the formation of today's land masses are considered. Earth's surface crust is as mobile as life; neither are immutable and at the close of the Palaeozoic era around 250 million years ago the planet's continents drifted together to form a single ocean and a solitary area of land, the super-continent of Pangaea. In time, Pangaea once more began to slowly fragment until, during the Cretaceous, a major breakup occurred. New continents formed and gradually groaned their way apart. Initially, one part of Pangaea (Gondwanaland) moved to the south while another (Laurasia) travelled north, eventually forming Asia, Europe and North America. Imagine a jigsaw puzzle floating on water, with the pieces slowly separating, their shapes eroding so they could never fit perfectly together again. A few centimetres a year, the plates still move in their unhurried, continental waltz. Perhaps in distant time our modern-day continents will again collide to reform a future Pangaea, in the same manner in which they coalesced the last time.

Where the plates collide there are opportunities for new mountain ranges, for example where the sea-laden Nazca Plate grinds its way beneath the South American Plate to create, centimetre by centimetre, the Andes and a chain of active volcanoes. Such boundaries create ideal situations for dividing populations, and none more so than when Pangaea fractured and water separated the land masses. Pangaea's animal life, at the time consisting of mammal-like reptiles, divided with it, some to drift through regions of polar cold while others were subjected to tropical heat. Gondwanaland fragmented only 160 million years in the past, and again split the life-forms it carried to evolve in their separate ways, driven by this changing environment. A splinter mass moved northwards from Gondwanaland to compellingly collide with Asia and form today's Indian sub-continent, at the same time slowly forcing up the Himalaya mountain barrier.

By inspecting the modern distribution of life on earth, we can see the effects of this ancient cause. Land areas, even on Gondwanaland, were not fixed; floods covered new habitats, and forced species to adapt and adapt again. The marsupials, the pouched mammals such as the kangaroo, evolved and spread. However, probably initially in what is now Asia, the placental mammals then arose. Compared with marsupials, placental mammals take their young further down the road to maturity before birth takes place, and this presumably gave them a crucial advantage when the groups competed. As the placentals occupied their niche, most marsupial species went into decline to leave only a fossil reminder in the rocks.

Except, that is, for one area. When Gondwanaland divided, the continental plate which was to become Australia isolated the marsupials, allowing them to evolve without interference from the placentals. It was only elsewhere that the placental mammals could compete; there was no land bridge they could traverse to reach their pouched cousins. Australia bears little fossil evidence of placental mammals ever having been present (other than bats, mammals which are mobile enough to have crossed ocean voids). In Australia, therefore, these castaway marsupials held sway over their domain.

Doria's Tree Kangaroo lives in the rainforests of New Guinea. As there are no monkeys, this marsupial (the joey's head can be seen protruding from its mother's pouch) has become arboreal.

So, competition is as great an influence on biodiversity as a changing climate. While massive changes, due to land masses inexorably carrying life on a slow, global cruise, aid the creation of new niches and provide the forces which mould organisms to them, anything which stresses a community – by removing or changing some element of it – leads to profound change. When a flood encroaches on a woodland floor, its inhabitants must leave for other areas where they create greater competition for resources. If a lone tree is struck by lightning, its food production ceases but shelter for animals, in the form of burrows or split bark, increases. The balance has altered, and some species benefit while others suffer: every cloud has a silver lining. Perhaps the changes force competing species to diverge: each requires the same food, but one now catches the prey in the open while the other alters its behaviour to ensnare its food in a burrow. Elsewhere, cooperation enables evolutionary advance: algae within corals, bacteria in the gut of mammals, interlinked life cycles to mutual benefit. Whatever the cause, changing goalposts proffer the raw elemental forces of evolutionary change.

All of which poses another question. Some habitats (for example, deserts) appear less rich than others (reefs, forests): are there some situations which favour the formation of a greater variety of life-forms, while others do not? Why should diversity have increased and increased yet again over the millennia, rather than stabilized at some undefined number?

If it is a changing environment, both physical and biological, which pressurizes species into diversification through evolution, it would seem reasonable that the areas with the greatest change will produce the most species. Unfortunately, any such assumption is invalid. A racing driver may be told that winning a race is simple: drive faster, then faster still. Although this is true, the pressures of increasing speed eventually become too great and the car inevitably crashes. If the speed of environmental change is too rapid, the process of evolution fails for many species. Those forms of life which are unable to adapt to the speed of change are doomed.

For species to have evolved, therefore, habitats must have changed at a viable, not too fast, rate. Once a community of species was formed, competition provoked (rather than restricted) the creation of biodiversity. Key species such as the starfish prevented any one life-form from becoming a despot; no one species could swamp its companions. However, observation tells us that some areas of the earth are richer than others: as all life evolved from common ancestors, there must be a reason for the unequal distribution of species.

To gather clues to why some areas of the earth support a greater biodiversity than others, it is useful to define what those areas are. It is then possible to extract any common features which might exist, such as climate or soil type. Biologists divide the world into areas based on the type of ecosystem; on land these major divisions are termed biomes. These biogeographical areas, linked to climate, are defined by some easily recognizable life-form which is widespread throughout the area, such as grass which defines a temperate grassland biome. Other areas include tropical rainforest, desert, mountains and tundra. Similar areas exist in the ocean, such as the deep, benthic sea bottom, estuaries with their brackish water, mangrove swamps and coral reefs.

Coral reefs may be the oldest of our ecosystems. The ability of corals to extract the raw elements of calcareous growth from tropical salt water and build huge, protective homes for their soft bodies is one which extends back to the first explosion of life which occurred in the Cambrian period, over 500 million years ago.

Corals, like many sea anemones, are symbionts: they hold within their coelenterate bodies small algae, called zooxanthellae, which thrive in sunlight to provide sugars to their sheltering coral cells while the coral provides the algae with a home. For this reason absolutely clear water is required to permit the sun's rays to penetrate, and even then the living portion of the reef is confined to the top few metres. Poor or variable quality water spells a death knell for a reef, where too much natural silt is enough to remove the ability to survive.

Such good conditions are scarce: ocean margins which are shallow, warm and clear, with sufficient sunshine to penetrate the water and sustain the zooxanthellae, limit reefs to an area of about 600,000 square kilometres in a tropical ring no more than 30 degrees either side of the equator. On the face of it, a mature reef provides perhaps the most stable of planet

Fallen trunks do not merely signify the death of a tree. When a tree dies, it opens an area of forest floor to sunlight, encouraging growth. As it decays, its nutrients are returned to the soil, and homes are provided for other species.

earth's ecosystems: water temperature and salinity normally vary little and, as nutrients are consumed, more are readily supplied. However, for the reef community to have evolved some changes must have occurred, and these pressures are readily supplied by factors such as hurricanes, disease, changes in ocean levels, and predation. In a forest a fallen tree provides change: a clearing, locally increased sunlight, nutrients as it rots, and the chance for new organisms to arise as they compete for life. There is little difference between this and a reef: no ecosystem can maintain a high biodiversity and also remain in total equilibrium.

Rainforests form a biome which staggers the imagination. Here are vast tracts of land bearing magnificent trees and an astonishing web of life. Giant plants soar forty metres or more into the sky, maintaining a canopy of leaves which is so dense that it blocks virtually all light from the forest floor. It is here, in the canopy, where most growth occurs and the major part of the food web is maintained.

Traditionally, rainforests (and their ocean equivalent, the coral reefs) have been considered the most species-rich areas of earth. Rainforests are also commonly believed to have high growth rates and productivity.

Productivity is defined in terms of the amount of growth, measured by weight. Biologists refer to this measurement as biomass: the biological mass. To find the biomass of a field, for example, take all the living material from one square metre (grass and other leaves and roots; worms, insects and bacteria), dry it, weigh it, and multiply the answer to match the area of the field. If the measurement is repeated after an interval of time, the difference in biomass indicates the rate of growth. This example would not yield a totally accurate result, for no account is taken of transient feeders such as birds, but the principle is clear. A measure of growth using biomass is more accurate than counting individual animals and plants or measuring height or new leaves.

Is there, then, a relationship between biomass and biodiversity?

A high biomass within an area is not necessarily the same thing as having a large number of species, though it would appear so in rainforests. Plankton drifting in the ocean possess a high productivity but a relatively low biomass: growth is fast and productivity is good, but the individuals soon die and return their nutrients to the sea, their combined mass never attaining great heights. A low biomass is also found on limestone terrains, but species numbers are high. In contrast, high biomasses exist in monocultures, where few species are found: farmed fields and mangrove swamps, both high production areas, are examples. The biomass therefore gives us no link to biodiversity.

But the nutrient supply which sustains growth is another matter, though with an unexpected conclusion. Consider where the majority of minerals are stored in a rainforest: in the existing woody biomass of the plants and leaves, in dead leaf litter, and the soil. Over 75 per cent of these minerals are locked in the wood; in other forests the majority of minerals are more normally confined to the soil. South American rainforests are found growing in ancient soils, over 100 million years old, and many nutrients have been leached away by the same high rainfall which supports the system. The abundance of life-giving water is the same factor which causes limitations, and nutrients are another resource for species to compete for. The fate of the forest's minerals indicates a desperate need to avoid the consequences of logging, because if the trees are removed the minerals depart with them. The poor supplies of trace elements left in the soil, bereft of tree protection, can now truly be washed away.

Coral reefs require very clean water for sustained growth, because sediment or water cloudiness will block the sun's rays and lead to bleaching – the death of the coral's companion algae, the zooxanthellae. Corals are efficient at extracting nutrients from seawater, but require the algae for survival as these supply them with sugars from photosynthesis. However, corals cannot cope if nutrients are present in excess; it is not like taking what is required from a supermarket shelf and leaving the rest for another day. When extra phosphates and nitrates are supplied, thread-like strands of

The Great Barrier Reef, off Queensland's coast, consists of a series of separate reefs and islands and stretches for a distance of over 2,000 km. The hard corals which produce reefs, perhaps planet earth's oldest ecosystems, require clean water and a high intensity of sunlight to support their companion algae. For this reason, they are restricted to the upper levels of tropical waters.

slimy, choking green algal growth overtakes the coral and it is forced into decline and death. The nutrients are not poisonous, but coral growth is slow and cannot compete.

The same pattern is repeated elsewhere: infertile places can support an abundance of life, as long as sufficient water is present. Heathlands are surprisingly rich in species' diversity, for example. One technique adopted by plants in areas of low nutrient levels is to turn carnivorous, capturing insects to obtain the nitrates and phosphates which are lacking in the sandy soil. Other plants may become parasites; unwilling to wait for the chance of growth on dead and decaying material, they attack the living. These and other environmental pressures have, seemingly, aided the formation of a diverse community.

Clear-cutting forests leaves a trail of destruction, opening the soil to the forces of erosion.

Another link can be forged between infertile ground and the variety and form of life. All species must balance their energy (and nutrient) requirements against activity and growth: one cannot be permitted to outstrip the other. Take an area where there is an abundance of nutrients and there is essentially no restriction on growth. Here, if leaves are eaten by animals, more can easily be grown and it may be a better use of energy to sacrifice the leaves.

However, plants growing in infertile ground find that the nutrients required to grow new leaves are at a premium, so some plants develop toxins as protection. Eucalyptus trees produce toxins only when growing on infertile soil; where growth is easy, lost leaves to herbivorous marsupials makes little difference but, where growth is slow and difficult, toxins preserve what little the plant produces. Plant diversity is enriched, but larger animals are not supported. Here may lie another cause of biodiversity: grazing in nutrient-poor areas forces plants to evolve defences. Then, as evolution advances, specialist feeders evolve to counteract these poisons, and the plants must diversify and . . . . Monarch butterflies, noted for their annual migration of several thousand kilometres between North and Central America, rely on milkweed plants for the caterpillar stage

Opposite: Eucalyptus trees can grow in a range of conditions, including poor, infertile soil. When nutrients are scarce, the tree is able to produce toxins as a protection against being eaten. However, where nutrients are plentiful, the energy involved in manufacturing poisons is too great, and it becomes more energy efficient for the tree to allow its leaves to be eaten.

Koalas are marsupials, and are now only found living in the eucalyptus forests of eastern Australia. They have unusual digestive systems which can cope with the vagaries of eucalyptus chemicals, and they are said to balance the effects of one leaf against another by feeding on different trees.

of their life cycle. Milkweed is poisonous, but the caterpillar can detoxify it, deriving a virtually exclusive food supply.

Even though a poor nutrient level can help initiate species' diversity by becoming a limiting factor, something seen in rainforests and reefs, this does not necessarily indicate poor productivity: rainforest species grow fast, although the increase in total biomass remains slight (adding a leaf to a mighty tree does not appreciably increase the biomass). Where there is good productivity, as in rainforests, in turn there is plenty of food to support a rich community of herbivores and their predators. The presence of animals, then, is likely to be highest in productive areas. With this further impetus comes more competition, and diversity increases once again.

These factors cannot be the end of the story, however. If nutrient supply and good production levels were the only things involved, there would be far more widespread, rich areas of life. Both deserts and the polar caps are relatively infertile yet, while they possess life, they are not as teeming as reefs or forests. Perhaps the extremes of temperature involved in desert and ice caps provide a clue.

If the number of species so far identified are plotted geographically on a map, it is readily apparent that they are more numerous near the equator and diminish towards the poles. It is even more noticeable that they are clustered in certain areas. Man, not nature, operates with government-controlled borders; species recognize only geographical and ecosystem barriers. While the economies of the Western world produce over half of the world's gross national products, over half of the world's known species are found in seven countries: Australia, Brazil, Colombia, Indonesia, Madagascar, Mexico and Zaïre.

Why here? For some species, it is due to a geographical isolation: life in Australia and Madagascar, after all, evolved without placental or other intrusive competition and produced a wide range of species found nowhere else. Mexico and Indonesia benefit from disparate but now coexisting species colonizing the land from two directions, the north and south. In Brazil, Colombia and Zaïre are rainforests, though the latter only possesses the remnants of a once proud woodland.

Of course, these ecosystems also exist in other countries of the world, but to consider only seven countries demonstrates how concentrated life can be. If using artificial, man-made geographical lines on a map is undesirable as an illustration, consider instead the rainforests and reefs alone. Even though rainforests only cover 6 per cent of the land surface and reefs less than 0.2 per cent of the ocean, well over half of earth's known species may be counted in residence. In fact, over half are located in the rainforests alone.

What is so special or unusual, apart from nutrient levels, about these habitats? Using the diversity–latitude link, and noting the tropical influence, what factors remain that might influence the spread of life?

When damp clothes are put on a washing line, several factors combine to affect the speed of drying: humidity, wind and heat. The fastest drying occurs on a dry, windy, hot day, while a foggy, still, cold day will have produced little effect by the time the washing is taken in. The same factors apply to plants and the process of transpiration (the process which moves water through the plant). A hot, dry, windy day will increase the water flow through a plant.

Transpiration is a logistical necessity. Leaves require sunlight energy to photosynthesize, so by default must be wide, flat and adapted to absorb the sun's rays. This also means they will get hot, so the leaf permits water to evaporate to cause cooling, and the transpiration stream must replace the lost water. You can observe the same cooling sensation if you spill petrol or nail varnish remover (acetone) on your hand; while the volatile liquid evaporates your skin feels colder as heat is removed. Sweating follows the same principle; evaporation of liquid from the skin's surface causes cooling, although it may not feel like this at the time because the sensitive nerve endings in the skin can only detect the locally increased temperature rise caused by increased blood flow near the surface. Transpiration is therefore driven by evaporation, which occurs because of the leaves' shape. This evaporation causes more water to be drawn from the roots, supplying more water for cooling and photosynthesis, as well as transporting minerals – another of its major functions.

The system only operates if there is sufficient water to restore any that is lost; leaves will inevitably, due to their structure, lose water and wilt if it cannot be replaced. This is the principal cause of leaf-fall in autumn – it is the only alternative to the water-deficient conditions which prevail in the winter, when water is locked away as ice and snow; rainforests, however, can maintain their leaf cover all the year round in their warm, humid environment.

Indeed, plants have developed a number of survival mechanisms to cope with the drying effects of warm weather. For example, leaves which appear shiny (rubber plants, holly) are coated with wax to retain moisture. Where water is truly at a premium, in deserts, leaves are modified to become dry, protective spines and cannot lose water. Others are covered in hairs to reduce the effects of wind, and small pores in the leaf surface (stomata) can open and close to control the rate of water loss. Coniferous (cone-bearing) trees cope with life at higher, cooler latitudes than the tropics while retaining their leaves all year round by reducing their leaves to needles; in direct sunlight large, flat leaves may attain a temperature some 15°C above needle-like leaves. With attempts to balance the rate of water loss against the requirements of cooling and liquid replacement, water availability is a crucial factor in the formation of rainforests.

Rainforest is defined as having a minimum of 100 cm of rainfall a year (over half the water evaporated back into the air through transpiration falls back as more rain, which consistently exceeds 200 cm of rain a year in South American forests). In contrast, places such as the Atacama Desert in Chile receive no rainfall at all, and plants here and in similar situations, such as the Namib Desert in south-western Africa, rely on mist and fog for moisture and store what water they can in swollen stems or roots (deserts are characterized by less than 25 cm of rain a year). Water is an essential pre-requisite of life; its presence, with high levels of sunlight supplying warmth, adds another factor to our biodiversity-causation list.

Owing to their design, leaves must lose water by evaporation to remain cool. In desert-dwelling plants, such as cacti, leaves have evolved to become spines. This gives the advantage of added water control; water is retained in the stem, affording a layer of protection against being eaten by animals.

For other reasons which might indicate why the tropics produce more species and more biodiverse communities than elsewhere, we also need to look again at the effects of geological time. Our land masses were still changing, then as now, and when the first corals began to appear, even before the time of Pangaea, today's Sahara was at the South Pole, locked in ice. Pangaea and its subsequent fragmentation is part of this possible explanation; as our modern continents formed, slowly drifting the globe, some passed through areas of hostile climate while others were more favoured. In the Carboniferous period Britain, then still part of Pangaea, was equatorial, only heading north after the divide. When Gondwanaland fragmented, what became Antarctica moved to the south while South America drifted through the tropics. The planet's climate, generally warm and wet, slowly became cooler and drier. Life in Antarctica diminished as ice gripped the land, and glaciers rolled from the poles.

Rainforests would originally have been spread over wider areas, but ice ages took their toll. What remained were the final refuges (known to biologists as refugia), where a wetter, warmer climate permitted survival. As conditions improved, the refugia formed a nucleus which could repopulate surrounding land, but only after life had continued its inexorable roll onwards through time and evolution.

Effectively, then, the maintenance of a relatively stable climate permitted evolution to continue in some areas for longer than in others. The refugia were essentially island communities and, in isolation, life diversified into newer species which adapted to details of their local environment. It is also possible that the designation 'island community' is a literal one. Over the past 100 million years or so, sea levels were much higher than they are today, covering low-lying land and separating the refugia from other remnant forests. As the polar caps froze, sea levels dropped and land bridges appeared; South America was linked to North America a mere 5 million years ago. In a less major way this cycle may have been repeated several times, isolating areas then permitting a remixing of new species before reisolating them again, each time the luxurious rainforest refugia maintaining their aloof, untouched position.

It may be, then, that the rainforests are so rich in life due to the simple factor that they have been permitted to evolve for a long, undisturbed time. With a stable climate, there have been fewer, major disruptions to the long chain leading from ancestor to descendant and, again due to this stability, there has been every opportunity for species to locate and enter new niches. When the niches are all full, further specialization – and speciation – causes the creation of new ones. Then, with a greater number of species and an increasing gene pool, the resources of life can increase yet again. In all this, rainforests offer an ideal environment filled with an astounding number of niches.

As an example, consider what is under your feet at this moment: carpet, soil, grass, it does not matter. You are oblivious to the precise nature of what you stand on, but there is life there nevertheless. You are living in your habitat, but below your feet are other niches too small for you to enter. Perhaps there are mites, or worms; there are certainly bacteria. Even if the ground is sterile, you carry life upon the soles of your feet, nestling in minute cracks and pores in the skin. You contain a rich fauna within your mouth and intestine; single-celled animals wander over your teeth, completing the digestion of remnant food. Your own body is sufficient to support a complete ecosystem. In 1892 in a popular science book, *Winners in Life's Race*, Arabella Buckley wrote:

> *We found creatures enough to stock the world over and over again with abundant life, so that even if the octopus had remained the monarch of the sea, and the tiny ant the most intelligent ruler on the land, there would have been no barren space, no uninhabited tracts, except those burning deserts and frozen peaks where life can scarcely exist.*

In the last, Arabella was wrong, for even in arid sands and the icy poles there is abundant life: the Antarctic ecosystem, in particular, is known to possess a rich, biodiverse collection of bacterial and algal species. Within any one habitat, large or small, there will always be many niches:

> *Big fleas have little fleas upon their backs to bite 'em,*
>
> *While little fleas have lesser fleas, and so ad infinitum.*

A rainforest is no different, but it is a vast place to imagine. In particular, there is a tower block of life which depends on the sun's rays filtering through the upper canopy. As elsewhere, there is a struggle between species: to gain an advantage over the competition is all important. A few taller trees protrude above the main canopy to receive their payment of extra sunlight, while those below are placed more into shade. There is always a negative side, however: the exposed leaves receive less protection from wind and, as transpiration and water losses increase, both leaves and trunk alike must become modified to cope with the increased stresses.

Down below, dimmer light requires larger leaves to absorb every available photon, but these can be more delicate as draughts are fewer and, in the humid air, drying out is less of a problem. Light diminishes along a gradient of conditions towards the twilight forest floor, where scavengers work on fallen wastes. This type of gradient is a common one in biodiverse areas. In a rainforest it is caused by light, while in the ocean depths there is a decrease in both temperature and light level, while water pressure increases. As a mountain is ascended, ultra-violet radiation intensifies and temperature drops.

Effectively, as with the niches beneath your feet, there are many situations which species can infiltrate, and where there is a gradient of conditions there are more niches than elsewhere. If a niche appears to be filled, there is always a chance for another organism to use it at an alternative time of year or during a different part of its life cycle. A hole in bark, a crack in rock, a leaf which consistently holds a crucial thimble of water near its stem: all can be exploited. Then, the next stage follows: with the hole in the bark and the dribble of water now occupied, there is an opportunity for a new predator or parasite to evolve. The more resources there are, the more niches can be created. As they become filled the competition between species increases yet further, new niches must be found, and the cycle continues. Effectively, this tends towards the creation of smaller and smaller niches, which perhaps explains the reason why there are fewer huge species than tiny ones, and fewer tiny species than minute ones. The outcome is an increasing number of adaptations and, through natural selection, new species. The outcome is one of greater biodiversity.

The specific answer to what factors lead to these most biodiverse areas of earth is difficult to ascertain with certainty, but it must encompass the concepts of evolutionary time, climate, solar energy, the size and isolation of communities, and available resources. Are any of greater importance than others?

Good sunlight, driving transpiration, leads to a greater number of species. It also appears that species' richness is linked with geographical separation, but that today's biodiversity depends less on this than a large enough area to diversify in, and a stable climate. If growth (productivity) rates are high enough, there is every chance for herbivores to establish themselves and, in turn, support a flourishing web of life. Plants adapt to cope with their attackers and, in turn, animals evolve to sidestep their food's defences. Deals are struck between species, with payment of nutrients or accommodation on offer in return for other nutrient supplies or protection. Species live together in harmony, or in postures of attack and defence. And every point of competition in this complex world leads to new, and more subtle, adaptations.

Neither must the role of change be ignored. Disrupt the environment in a sudden, devastating way and life does not survive. Disrupt it in a manner which organisms can adapt to, and life marches towards evolution. Over-predation reduces numbers of species by permitting one

to take over the system to the detriment of others, but predators (and disease and other variations in nature) in balance cause an increase. We need storms and seasons, fires and floods, to help maintain species richness. Make the ecosystem too stable, and speciation again falls. A realistically dynamic, natural world is one which maintains and expands its diversity.

However, much of this generalized hypothesis is difficult to check. As every scientist knows, a hypothesis must be tested by a repeatable, fair experiment before it is accepted as truth. At best, our answers are often in the form of 'this explanation fits the facts as we know them, and it's the best explanation we have'. Hypotheses and ideas are working tools, to use however we may choose.

Unfortunately, in the realm of ecological studies – and particularly with respect to such huge, far-reaching questions as 'where does biodiversity come from?' – valid experiments are far from easy. How do you viably take what you believe to be a crucial factor from one area of forest and compare it with another, unchanged area without destroying what you wish to study? We are left with minute experiments and extrapolated results to interpret. To predict the movements of one animal in a jungle is impossible, but to predict a trend for the species is viable. That is, given the data; when the problem encompasses all animals and plants in an area, plus the effects of terrain, resources and climate, we are dealing with almost infinite potential ramifications. That data is just not available; after all, though we are improving, we still cannot accurately predict the weather.

While twelve good men and true may be swayed by the court's arguments and tend towards a decision, they wonder what new evidence remains concealed from understanding and has yet to be revealed. This is an enigmatic earth and, as to the truth regarding biodiversity, the jury is still out.

Rainforests form inconceivably immense communities of life, relying ultimately upon water, sunlight and warmth to maintain their domain.

CHAPTER SEVEN

EXTINCTION IS SO FINAL

The stages of creation, except some rare oscillations, follow a rising scale. Nature seems to proceed by a succession of essays before fashioning her more splendid chefs-d'oeuvre.

The earth is only an immense cemetery where each generation acquires life at the expense of the débris of that which has just expired. But we have now reached an epoch of transition; the exhausted creative powers are experiencing almost a period of arrest; they are waiting till new telluric perturbations awaken them from their torpor!

F.A. Pouchet, *The Universe*, 1882

We have an image of life on earth, one of a riot of colourful biodiversity, spreading and coexisting in glorious harmony. There is a difficulty with this vision, which was the cosy one of a Victorian past when natural history was the considered work of the Creator. Humans are part of the 'harmonious' whole and, like the worst dictator, have developed formidable powers without the concomitant wisdom. With the effects of our actions, Life Burning Brighter (the title of Chapter 6) can have a literal meaning. Extinction is not something relegated to the distant past; biodiversity is (not might be, but *is*) crashing in flames. In extinction there is something more fundamental, more permanent, than death itself.

Our current, well-heralded, wave of extinction seems awesome, alarming, frightening. We can also look to our knowledge of other times when, as every schoolchild is aware, the fossil remains of dinosaurs tell a similar story. This age of the 'terrible lizards' precedes that of man, and it is not the only occasion when there has been a massive change on earth. Is there anything to learn from past extinctions which might help us understand the reasons for today's crises?

The fossil record is widely used to support Darwin's theories, though Darwin himself believed that it was only one element of the whole. The argument is that, if evolution consists of small changes which gradually lead via natural selection to the formation of new species, a series of stony images of once-living organisms should remain to tell the tale. In the oldest fossil layers at the base of an undisturbed cliff will be the species extant at that time; higher in the cliff are the younger layers, where sediment has again trapped, encapsulated and crystallized the species of that period. In between, locked in the earth, are layers which contain a succession of life, building up through ancestor–descendant chains.

However, what we also find are anomalies and gaps in the records, like a mystery novel missing a crucial chapter. While every species progresses through an evolutionary book, character by character, page by page, some stories stop abruptly while others progress to the present, but without documenting every step in the sequence. Some species become extinct, while others lack the minutiae which are required for a good story. Reading this novel is difficult; the central plot is understandable but, without the crucial details, deductions must be treated with care. How would the following sentence be completed without other information: 'The — went to the station.' The missing word or words might be anything: man, woman, man and woman, horse, train; in such situations, deductions are no more than guesses.

The information missing from the fossil record provided Darwin's opponents with one of their most important arguments: where were the links between older species and their descendants? As fortune had it, some of these fossils were subsequently discovered, but the record nevertheless remained, then as now, incomplete. Darwin acknowledged the problem, but felt that the formation of a fossil was a rare event. The rocks would therefore never hold a perfect representation of his vision; fossils studied in isolation would always yield anomalies.

A modern study of Lake Turkana in Kenya revealed a smooth progression of freshwater shellfish fossils, apart from two abrupt changes in the time line which suddenly produced new species at the expense of the old. Notably, these changes affected all the shellfish at the same time.

In an analogy which extends across the whole of life's record, it is like having a library of different books, all with some pages missing. The books represent species that came into being at different times, so some are long (an ancient species) while others are short. Some abruptly stop (the species has died out) while others are still being magically written; turn to the back page and watch, at an infinitesimal rate, new characters forming. The missing pages, say pages 45 to 46 in one volume, represent a gap in the tale for that species. At Lake Turkana, *every* book misses a single page at some time-consistent point. The story is not only incomplete, it is *consistently* incomplete. For one species' fossil series to be interrupted is entirely believable, but could this 'multiple species interruption evidence' point to evolution, now and then, going through a sudden period of creation? Perhaps evolution moves, not slowly by natural selection, but under another force that springs new life-forms into existence at paragraph-skipping intervals.

There is an alternative explanation for these judders in the fossil records: the rate of fossil production may be faulty, not the concept of a smooth gradation of change under the control of natural selection. To record the minutiae of evolution, the formation of fossils must be regular throughout time. This may not be the case. In our library analogy, the rate of sedimentation which buries the mollusc corpses may have been constant for the first few chapters of the books, then slowed to an imperceptible level (pages 45 to 46) before returning to 'normal'. To continue the analogy, the information was conceived (the 'missing link' shellfish existed) but it was never bound into the volume (the fossils never formed). It is not so much that the pages never existed, but that the bookbinder never included them. The text appears to jump, although the overall plot is still understandable.

Yet, while the explanation is plausible and logical, it appears more likely that in this specific case some environmental change suddenly pressurized the shellfish to adapt. A study of non-mollusc species at Lake Turkana indicates no comparative jump in the fossil story; they were not similarly affected. Whatever changed, it affected only one group of organisms; the evidence supports a fluctuation in the water level of the lake, indicating that the shellfish did, indeed, suddenly (rather than gradually) move up the evolutionary tree under the pressure of a changing environment.

Instead of challenging the theory of slow, evolutionary change, it is the speed of adaptation which the facts appear to dispute. Two hundred generations might, biologically, be enough for the formation of a new species, yet this would be as nothing to the juggernaut of geological time. The minute changes in two hundred generations of progeny would not show up in compressed rock strata: the fossils would simply spring from one end of the sequence to the other. The Lake Turkana record, like others, can therefore be explained using natural selection (in this case, acting upon variations suited to changed water level) coupled with an understanding of fossil production. Equally, just as environmental changes caused a spurt in evolution, they could equally have wiped out the molluscs altogether; there is a fine line between what life can cope with, and what it cannot. What promotes the production of one new species will often destroy another.

Today's crocodiles, lizards, turtles and tortoises are descendants from the great age of the dinosaurs. As reptiles, they are cold-blooded, so must obtain sufficient heat from their surroundings to permit movement. This limits crocodiles to warm climates, where they can swim and run with surprising speed. In the water they are efficient predators, even to the extent of swallowing up to 5 kg of stones to act as stabilizing ballast.

Throughout the records that the rocks conceal, there are spikes of disruption (at least ten; estimates place these at roughly every 26 million years) which made serious inroads into our past's biodiversity, with five representing major catastrophes. These occurred at the end of the Ordovician period (440 million years ago), the Devonian (360 million years ago), Permian (250 million years ago), Triassic (200 million years ago) and Cretaceous (65 million years ago); indeed, it is the changes in rocks which represent these mass extinctions that led to geological time being sliced into its many-named segments. In each case there was more than a unified gap in the fossil record which required a missing link for single species: there were mass extinctions as life-forms disappeared from the earth, for ever.

As with the Lake Turkana explanation in which the fossil record indicates a jump in the development of a species, the common inference is that, on each occasion of a mass extinction, something suddenly happened to wipe out life and promote the production of new species to take its place: evolution, it seems, works fastest during times of maximum stress. Again, the concept of 'suddenly' requires examination. Geological time covers unimaginable numbers of years: the fossil evidence indicates that dinosaurs did not disappear overnight, as popular myth would have it. Nor are all the ancient animals which became extinct lumped together in one, classifiable group: far more species than the dinosaurs disappeared from the earth at that time. In fact, some losses seem to be spread over millions of years; on the fossil evidence, dinosaur species were in decline for 10 million years before something finally eradicated them. Even then, not all their relatives were eradicated. In the

Triassic period, crocodiles belonged to the archosaur group of reptiles, which gave rise to the Cretaceous dinosaurs. Crocodiles moved through time to become the survivors of the Mesozoic era, which closed alongside the age of dinosaurs at the end of the Cretaceous, and their descendants still roam the warm waters and muds of river banks today.

Even so, while the dinosaurs may not have represented a 'here today, gone tomorrow' story, mass extinctions, as evidenced by multiple species' loss through a common period of time, have nevertheless occurred. These extinctions lead, of course, to changing pressures on remaining species, and are therefore an integral part of the natural process of evolution itself. While the source of change which caused these extinctions remains conjectural, a series of hypotheses do exist.

Whatever the changes were, it is certain that they must have been significant to have initiated such widespread effects. The commonest ideas concern the effects of changing climate or vegetation: a cooling earth, for example due to a glaciation which covered the land during an ice age, could have affected plants to the extent that the dinosaurs had insufficient food to survive. Such conjectures beget the question: what might have caused any such catastrophic alteration in climate? Might some other factor be involved? The events causing mass extinctions were certainly catastrophic, for the inhabitants of our seemingly rich earth of today represent less than 1 per cent of the species which have existed in the past. Using this estimate, based on species' losses due to mass extinctions and the background rate of extinction (the rate of species' loss which occurs day by day), our planet has already lost virtually all the life-forms it once possessed.

The end of the age of dinosaurs is perhaps the best studied of the mass extinctions, probably due to a morbid fascination in the ways that such a magnificent dynasty might have ceased to exist. The Cretaceous community of 65 million years ago included the dinosaur species now familiar from their reconstructed fossil skeletons, as well as the ocean's plesiosaurs and ammonites (cousins of today's octopus), and the pterosaurs of the air. All disappeared virtually in totality, to leave the birds and mammals behind. In this one event about three-quarters of earth's species became extinct.

Lufengosaurus, one of the earliest dinosaurs, lived in the Triassic period; this specimen was found in China and stood about 6 m tall. The fossil remains of such animals tell us much about their diet and behaviour, but the reason for their demise is still a matter of conjecture.

Although the fossil record is poor, and there are indications of a declining diversity of dinosaurs well before the end of their era, a sudden catastrophe which decimated life on earth is nevertheless indicated. A study of fossil pollen shows that plant extinctions were more extreme in North America than New Zealand, yet those extinctions which took place seem to have all occurred at the same time.

Temperature changes are often cited as a contributory factor, slowly altering the earth's climate in a way which species could not adapt to. However, the plants which were extinguished have modern-day cousins which are able to survive in chilly conditions. Work in Australia and Alaska indicates that dinosaurs were able to fare well in the cold, living in environments where the temperature never rose above 12°C.

If slowly increasing cold was involved in this mass extinction, such as that caused by polar weather encroaching on successively lower latitudes, the extermination of plants would be spread through a greater time period. As for the geographical difference, perhaps some sudden catastrophe abruptly affected the northern hemisphere in the spring, while, in the south, autumn plants had already prepared for winter. In addition, the inferred existence of cold-surviving dinosaurs challenges the concept of temperature being a factor at all, unless it was part of a *sudden* catastrophe which also affected other components of the environment. For example, a steady decrease in heat might cause sudden plant extinctions as successive species passed a threshold and succumbed to cold, but this would not explain differences between extinctions on different land masses at the same latitude. The pointers are there for a single trigger being pulled on a planetary-wide gun. We can only puzzle over what we learn, and seek further clues: what on earth might have caused such a swift, mass extinction?

'What on earth?' might be a misnomer, for there is evidence to support the idea that this most recent of losses was triggered from the heavens. Scientists have made studies of minerals occurring at the junction of the Cretaceous and Tertiary rock layers. This is usually termed the K/T boundary, taking the letters representing the names of the geological periods (K is used to represent the Cretaceous period, from the German word *Kreide*, meaning chalk). At the K/T boundary is an unusual layer of iridium: the muds and clays contain a high percentage of the rare metal, but it is not found elsewhere in the layers of sedimentary rocks.

Iridium is not common on earth's surface, but metallic meteorites contain high proportions of the hard, white element. In 1980 a group of scientists put forward a theory that it was a massive collision – a meteor, comet or asteroid strike – which caused the loss of the dinosaurs, and the search was on to prove them right or wrong. Controversy continues to this day, but supporting evidence for their idea was indeed found and the theory has become more acceptable in some quarters.

In any such collision, the theory stated, there would be a huge crater formed. Also, while the impact would have conceivably vaporized the meteorite and spread its dust around the earth to settle in the half a centimetre of muds of the K/T boundary, there would presumably be localized damage that remains to be found. Geological examinations revealed that such a place existed, near the coast of Mexico's Yucatán peninsula. Here, quartz grains have been stressed and broken into fragments (something attributable to an incredibly powerful shock wave) and a 50 cm thick layer of the glassy remains of molten rock is found.

Whether it was here or at another site, it is estimated that a meteorite about 10 km in diameter struck the earth at a closing speed of over 70,000 km per hour. That's a colossal force: the meteorite would have traversed the distance between the earth and its moon in only five hours. In comparison, Concorde's flight from London to New York takes three hours and fifty minutes: the hypothesized lump of rock could cover this distance in only six minutes, though without the same requirement to slow down at the end.

In such an impact, the force released would create earthquakes and throw up 1 km high

Opposite: The volcanic crater of Mount St Helens, Washington State. The blast in 1980 stripped hillsides of trees and affected living organisms 400 km away. In this picture, taken thirteen years later, plants are beginning to recolonize the bare ground between the skeletal remains of the former forest.

tidal waves; fossils are jumbled in the Yucatán area, land organisms mixed with those of the sea as if churned by water before being buried under sediments. Shock waves would have raced the earth at ballistic speeds, and flames would have rolled over the land. Dust thrown into the air, carrying the iridium, might easily circle both hemispheres of the globe from this tropical Yucatán site. Acid rains fell, washing nature's pollution to the land; where the rocks preserving the K/T boundary have survived the subsequent aeons of weathering, this layer is astonishingly uniform. The sun was blotted out and, when the skies cleared, leaves were coated in sticky dust so that the life-giving chain of energy which drives photosynthesis was destroyed. All in all, this unimaginable impact – as a measure of the energy released, exploding all earth's nuclear weapons at once, in one place, pales into insignificance – brought worldwide devastation within the space of days.

There are still difficulties in a full acceptance of the meteorite theory: the dinosaurs were already in decline at the time of the impact (why?), and fossils show that some survived beyond the catastrophe. An alternative theory is that volcanic activity produced the layer of iridium (the metal is more common in the earth's core than its crust) as well as choking the air with poison gases and acid rain, and filling the atmosphere with ash. The localized effects of the 1980 Mount St Helens eruption in Washington State did just that. It flattened forests and tore plants from the soil, leaving a lunar landscape of desolation. Over 400 km away plants were coated in ash and died, unable to photosynthesize, while nearby the ash built up into a 75 cm thick layer.

Even so, on a global scale the spectacular eruption of Mount St Helens provided a mere blip in the world's climate. The island volcano of Krakatoa, between Sumatra and Java, erupted in 1680 and again on 27 August 1883. The devastating sound it produced circled the globe and reverberated back seven times; 40 m high tsunamis (tidal waves) rolled away to kill 36,000 people on nearby islands. The crater which remained, beneath the ocean, spanned 7 km. An even greater explosion was described by Élisée Reclus in 1868:

> *In 1815, Timboro, a volcano in the island of Sumbara, destroyed more men than the artillery of both the armies engaged on the battle-field of Waterloo. In the island of Sumatra, 550 miles to the west, the terrible explosion was heard, and, for a radius of 300 miles round the mountain, a thick cloud of ashes, which obscured the sun, made it dark like night even at noonday. This immense quantity of* débris, *the whole mass of which was, it is said, equivalent to thrice the bulk of Mont Blanc . . . fell over an area larger than that of Germany. The pumice-stone which floated in the sea was more than a yard in thickness, and it was with some difficulty that ships could make their way through it.*

Timboro's eruption was five times the force of Krakatoa and left a 65 km wide crater. Yet, while other, greater volcanic blows to the land have been dealt in the past, not one is linked to the massive loss of species that is attributed to the Yucatán strike. As shown by the limited, though devastating, localized effects of these events, a single volcano cannot account for a mass extinction: even Timboro, though it severely affected crops in New England the following year, failed to produce any lasting effects.

To create a mass extinction an almost unprecedented, worldwide volcanic activity would have been required, but we do know that the earth is not a solid, stable planet, any more than the life which it supports. Could volcanoes alone, rather than the controversial theory of extra-terrestrial causes, be sufficient to explain the occurrence of the widespread devastation which the facts point towards? It seems possible, though difficult to substantiate.

Even better might be a combination of the two theories. Any meteorite impact could, as well as creating its own devastating effects, have triggered volcanic activity, providing both camps with a possible solution: a sudden

This representation of the eruption of Krakatoa in 1883 shows some of its devastating force; six months later, the atmosphere was still affected in London, over 10,000 km away. Dust ejected from volcanoes may have a significant cooling effect on the planet. In 1990 a Philippine volcano, Mount Pinatubo, erupted and ejected 20 million tonnes of sulphur dioxide into the atmosphere, resulting in a worldwide average temperature drop of up to 0.5°C which lasted for two years. To place this in context, the average world temperature of the last ice age was only 4°C cooler than today. A series of volcanic eruptions could therefore have had a significant effect on climate at the close of the Cretaceous period.

collision followed by a prolonged assault from planet earth's interior. This, indeed, could have caused a mass extinction, though there is still widespread dissension. While the effects of a single meeting with an extra-terrestrial body, or a restricted period of volcanic eruption, are not necessarily enough to provide the required worldwide effects, the sudden combination could be crucial for biodiversity.

Before attempting to place numbers on species removed from the planet, it is essential to realize that such estimates are, at best, approximations. A major difficulty stems from the fossils upon which we base our theories; after all, our knowledge of what once lived is based on images set in stone, and the record is fully acknowledged as being incomplete. The conditions required to fossilize an organism are very restricted, and more likely to preserve hard body parts such as bones or shells than the many invertebrate, soft-bodied species. Some insects were encased in their entirety in amber, a fossilized, translucent resin produced by trees, but even here the completeness of the record is woeful. In other cases, it is difficult to identify different stages in a life cycle as belonging to the same species. As insects grow, they may alter their form: a maggot to a fly, a caterpillar to a moth. The same argument applies to the trilobites, which shed their exoskeletons and changed their shape as they grew.

Additionally, how do we determine what a species is from a fossil? The modern working definition of a species, albeit an imperfect one, is that it can interbreed with other individuals of the same species. With regard to fossils, there is no opportunity to test the criteria: no matter how long two fossils are left together in a drawer, there will never be any fossilette offspring!

Palaeontologists therefore use evidence of comparative anatomy and the separation of organisms through time to deduce whether or not two individuals comprise different species. Even this is imperfect, and it is easier to relate organisms to the level of genera or families (larger groups of organisms) than to species' level. However, using mathematical extrapolation, the number of species represented by the lost families can be estimated.

Something in excess of 250,000 species of fossil are known from all the geological ages combined, which means that the number of actual species – taking into account those which have not been identified or preserved in the first place – must be much higher, certainly by orders of magnitude. The estimates of total life on earth range from 5 billion to 50 billion species through all time, with about 60 per cent of these having been lost in mass extinctions. The remaining 40 per cent belong to the ongoing, natural 'background' rate of loss which occurs under the daily force of evolution.

At the time of the formation of the K/T boundary at least 75 per cent of earth's species are estimated to have been lost, but even this was not the most devastating mass extinction. The earliest major spasm, at the end of the Ordovician, removed many brachiopod and trilobite genera, an estimated 85 per cent of all species. A period of 80 million years of further evolution followed which rebuilt earth's biodiversity. By the end of the Devonian period earth possessed land-based plants, amphibians and insects, but the mass extinction which followed reduced the number of species once again. This time over 80 per cent of species disappeared, including the graptolites (marine invertebrates) and many forms of fish.

The Permian/Triassic boundary bears a message of biodiversity destruction which exceeds all others: the trilobites, which had been on earth since its earliest fossil records, were gone after a reign which had lasted nearly 300 million years. Thus, some 250 million years ago, over 90 per cent of all species (and over 80 per cent of the genera; the estimated upper limit of species' loss is a staggering 99 per cent extinction) failed to survive whatever upheaval earth's climate threw at them. Between each extinction the numbers of species climb again, only to be brought down with a crash. Only 50 million years after the Permian extinction, in a smaller extinction spasm, the mammal-like reptiles ended their existence in the Triassic. It seems as though the rules of life are fickle; at such times as the Permian extinction there have been close calls for continued life on earth.

If, for the moment, a combination of a meteor or comet strike and widespread volcanic activity is accepted for the most recent extinction at the end of the Cretaceous, what could have caused these earlier catastrophes? Massive climatic (or other) changes are indicated, or else the mass extinctions would not have occurred. Again, there are different theories, some of which seek to explain the presence of iridium by other means; discredit the K/T boundary explanation, and there is less evidence for a major collision with an extra-terrestrial body at any time in the past. One suggestion is that the iridium could have been concentrated at the boundary as shallow seas dried out, but this explanation has not met with much favour. On the other hand, perhaps earlier collisions have occurred with bodies which did not contain iridium, and initiated older mass extinctions. If so, to support the theory of a meteorite strike, where are *their* craters?

On land, huge craters exist though none are of the same size as that near the Yucatán coast: the Chicxulub crater is over 200 km in diameter. The next nearest contender is in Siberia where the Popigai crater is 100 km across, and others could be hidden within the depths of the oceans. The frequency of chance collisions with solid bodies from within the solar system is greater than might be thought; shooting stars represent the trails of small meteorites, burning up in earth's atmosphere. Larger sizes are common, fist-sized blocks falling to earth every few years. Of course, much larger objects than this are believed to be required to create any realistically devastating effects. However, to place the known craters (almost 200 of them) in context, one which is 10 km wide would represent some 10,000 megatons force – the equivalent of the world's entire collection of nuclear weapons detonated at once. The search for craters has not revealed the evidence of enough major collisions to account for the previous extinctions, so on that basis it seemed that massive strikes are not that common – but the story continues.

In 1908 hundreds of square kilometres of Siberian forest were flattened by the shock waves of an exploding object from space which never reached the ground. In 1972 a film was taken of a massive chunk of rock which raced through Argentinian airspace; it had narrowly missed the

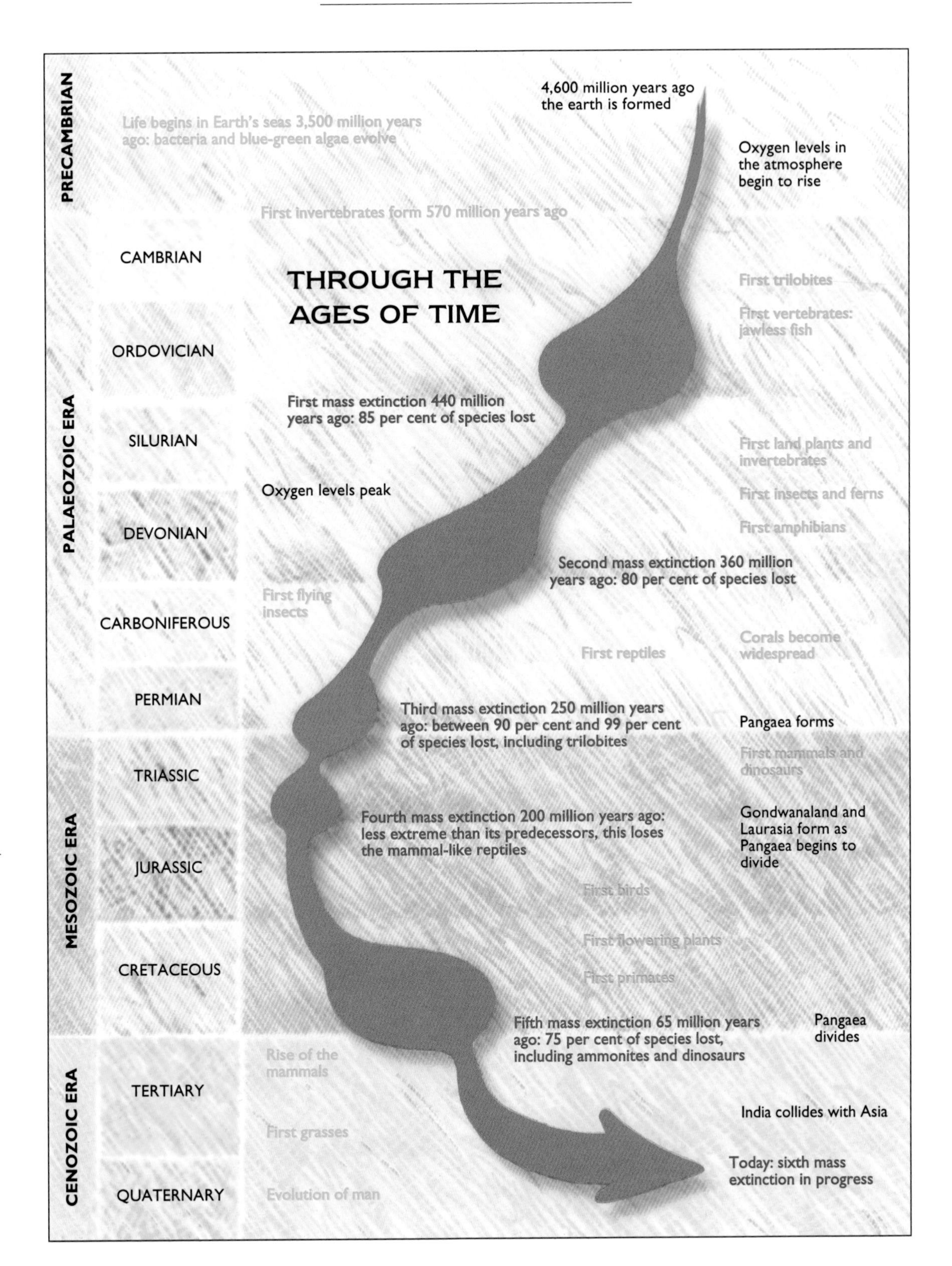
PRECAMBRIAN
PALAEOZOIC ERA
MESOZOIC ERA
CENOZOIC ERA
CAMBRIAN
ORDOVICIAN
SILURIAN
DEVONIAN
CARBONIFEROUS
PERMIAN
TRIASSIC
JURASSIC
CRETACEOUS
TERTIARY
QUATERNARY
THROUGH THE AGES OF TIME
4,600 million years ago the earth is formed
Life begins in Earth's seas 3,500 million years ago: bacteria and blue-green algae evolve
Oxygen levels in the atmosphere begin to rise
First invertebrates form 570 million years ago
First trilobites
First vertebrates: jawless fish
First mass extinction 440 million years ago: 85 per cent of species lost
First land plants and invertebrates
Oxygen levels peak
First insects and ferns
First amphibians
Second mass extinction 360 million years ago: 80 per cent of species lost
First flying insects
First reptiles
Corals become widespread
Third mass extinction 250 million years ago: between 90 per cent and 99 per cent of species lost, including trilobites
Pangaea forms
First mammals and dinosaurs
Fourth mass extinction 200 million years ago: less extreme than its predecessors, this loses the mammal-like reptiles
Gondwanaland and Laurasia form as Pangaea begins to divide
First birds
First flowering plants
First primates
Fifth mass extinction 65 million years ago: 75 per cent of species lost, including ammonites and dinosaurs
Pangaea divides
Rise of the mammals
India collides with Asia
First grasses
Today: sixth mass extinction in progress
Evolution of man

earth but, if it had struck, it is estimated that the force would have been equivalent to five Hiroshima bombs. Of even greater interest, in 1993 a comet passed too close to the planet Jupiter and was torn to shreds by its massive gravitational forces: science was able to observe the results. On 16 July 1994 the first fragment collided with Jupiter, leaving effects which extended for thousands of kilometres – enough to envelop an object the size of earth. A fireball, estimated at reaching 20,000°C, erupted into space before being dragged back to Jupiter. The outcome was astounding: the fragment created an effect far out of proportion to its size due, presumably, to its high density. And, if this could happen so coincidentally while mankind happens to be watching, could not such collisions be far more frequent than we assume? We do have evidence of sooty remnants locked in the K/T boundary to indicate a possible wildfire across the globe, although the evidence could equally have been left by volcanoes.

In the latter half of the 1980s a study was made of volcanic areas throughout the world, noting where floods of basalt – ancient regions where lava once flowed – were situated; a major one (the Deccan Traps) covers the western portion of India, for example. If a meteorite impact was great enough, especially if it hit an area of weakness, the earth's mantle could have been disturbed sufficiently to release lava and fill the crater. In addition, these solidified basaltic lakes are today linked via ocean ridges and mounts to 'volcanic hot spots' which are susceptible to eruptions, indicating their former activity during the relentless drift of the continents. Dating the lava upwellings (there are ten or more of them, with over forty major hot spots catalogued) produced, interestingly, a good correlation with the timing of the various mass extinctions.

Opposite: When volcanoes erupt, the lava they emit does not always form a basalt flow. If cooling is rapid, volcanic glass – obsidian – is produced. Unlike lava flows, which slowly weather and are eventually colonized by plants, obsidian rarely supports a living community.

The chain of proof is suspect, however: at intervals matching the mass extinctions, did a meteorite strike the earth and initiate the volcanic upwellings which conceal the extraterrestrial evidence, or did the crust weaken and release massive lava flows without any external influence? Could a comet have fragmented to strike the planet with a multiple, pepperpot effect (spattered craters are found in Australia which indicate this can occur), triggering a sequence of volcanic activity, or could the volcanoes have thundered into life of their own accord? Or were there different causes for each extinction, such as the fluctuations in sea levels which we know occurred 250 million years ago in the biggest of all the losses, at the junction of the Permian and Triassic periods?

When Pangaea formed it fused together the land masses of the Permian period, incidentally reducing the coastal shallows habitats which were so rich in life; unconnected or not, the trilobites and many other invertebrates disappeared. Crinoids, at that time a commonly occurring marine animal, all but disappeared, leaving only one genus to continue into the Triassic; corals and shelled animals were decimated. Land and sea were equally affected. We believe that the oceans receded, further reducing habitat areas when coastal shallows and marshes were left high and dry, leading to a loss of biodiversity. However, what was one organism's loss would be another's gain: the philosopher's stone, to science, is to determine some single factor, some irrefutable piece of evidence, which would help to explain the widespread, catastrophic loss of life.

Carbon, on which all life is based, occurs in different forms: carbon-12 and carbon-13 are relatively stable, while carbon-14 is not; it is the degradation of the carbon-14 isotope which is used in carbon dating techniques to determine the age of organic material. Because of its half-life, the latter isotope gives a limit on dating of about 40,000 years; beyond this, the amounts present are too small to be measured. However, the proportions of the 12 and 13 isotopes can be measured in limestone, itself formed from the remains of dead organisms. As carbon-12 is used in preference to carbon-13 for growth, it can be deduced that when there is little carbon-12 present in the rocks there must be good growth of organisms, locking the carbon into the world's biomass. When growth is poor, carbon-12 is returned to the environment and therefore shows up in a greater proportion in rocks. The data is useful as it can point towards periods of rapid and slow growth in the distant past.

We know that, during the Permian period, coal was formed from plant material, thus trapping carbon-12 atoms. However, from a study of Permian rocks we know that the period ended with a sudden upsurge of carbon-12 in the air: it must have been released either from coal, or as a result of the mass death of earth's organisms. There may be a clue to the Permian extinction here.

Back to the receding seas. While it could be argued that the mass extinction indicates a mass release of carbon back into the cooking pot, a reduction to fewer species does not necessarily indicate a lower biomass: surviving plants, for example, would continue to grow and expand into vacated niches. However, one theory looks to the carbon-rich coals and shales which were being exposed to air as coastal shallows drained. As these rocks eroded and broke down, and organisms died and rotted, two things would occur: the proportion of carbon-12 in the air would increase and, at the same time, carbon dioxide formation would remove up to half the oxygen. To proponents of this theory, the world's atmosphere then entered a period of global warming (the greenhouse effect) due to the rise in carbon dioxide and, as the polar ice caps melted, water returned at a massive rate (there is evidence for this occurring at tens of centimetres' rise in ocean levels per year), flooding the land once again while at the same time reducing the oxygen of the seas. In short, according to this scenario, terrestrial and aquatic life both suffocated.

So, no comet is needed for this theoretical explanation of extinction. But science also has other, optional, theories. Iridium has been touted as the ultimate indication of an astronomical strike; why not some object which does not contain iridium, but is rich in some other element? Nickel, present in high proportions in asteroids, has been proposed as it is extremely toxic to life in concentrations above 40 ppm; the fallout from a 10 km diameter, powdered asteroid would be more than enough to poison the earth, the nickel then leaching away without leaving a record in the rocks. It has even been suggested that the rise of the flowering plants, with their ability to manufacture bitter-tasting alkaloid poisons for defence, contributed to the loss of the dinosaurs: reptiles are notoriously insensitive to bitter tastes and will eat plants shunned by other creatures. Realistically, this seems an unlikely primary cause as even the carnivorous species died out, and these would surely have found enough food amongst the surviving species to continue along their evolutionary path. Another theory relates to bursts of neutrinos, produced by collapsing stars. Neutrinos (atomic particles) pass through living tissue relatively easily, but can cause damage to DNA. Comparing the effects of known radiation sources, calculations show that cosmic sources of neutrinos might cause significant degrees of cancer – especially in larger animals where there is more body mass for the particles to collide with. Could one of the mass extinctions be due to widespread cancers?

Of course, without proof of such occurrences scientists can only form hypotheses which are accepted or rejected; conjecture is easy and rife. In short, we do not know for certain what caused the massive population crashes of the past, but the search for evidence continues: it might be crucial to provide a yardstick for our own future. While we have no hard conclusions for the causes, we can hesitatingly state that there were, certainly, sudden mass extinctions which wiped away millions of years of evolution.

In the wake of destruction comes rebuilding. The chain of evolution depends on change: without change to create new pressures, there is no niche for a new species to adapt to. To have such massive extinctions simply opens up the opportunity for remaining species to rapidly diversify into untold, empty niches. In every mass extinction, the losses were soon outweighed by a new and splendid array of species: each loss was, ultimately, biodiversity's gain.

Supply an empty niche and life will radiate into it. The colonization of the land is one example. As plants began to grow in soil, they diversified through the empty habitats. Club mosses gave way to conifers and ferns and, where plants provided a new source of food, insects and other animals could follow, each species demanding its own ecological niche.

Then, as the Permian trilobites took their final bow, the ammonites rose to fill the earth's oceans and mammal-like reptiles became abundant. In turn the Triassic crash removed the mammal-like reptiles; insects spread and the mammals and dinosaurs arose. When the Cretaceous abandoned the dinosaurs, mammals were left to erupt across the earth, accompanied by life's recent invention, the flowering plants, and a few surviving dinosaur groups.

At every stage it was the evacuation of earth's habitat niches which helped permit this explosive radiation of evolving species: without empty niches to compete for, the power of natural selection would be constrained to find new (and almost by default, slower) ways of exploiting the environment to produce new life-forms. Add to this the slow revolution of continental plates through different latitudes and the creation of new habitats under a changing climate, and evolutionary pressures were indeed strong through the ages gone.

Ammonites, such as these examples of *Promicroceras planicosta* from the early Jurassic period, were cephalopods and were therefore relatives of the octopus and squid. One theory of their demise is linked to the rise of strong fish and reptilian predators which could crush the shells. Whatever the reason, ammonites disappeared from the earth, together with most of the dinosaurs and over 50 per cent of all marine invertebrates, at the end of the Cretaceous period. They had lived on earth for 325 million years.

As each mass extinction brought species tumbling down, some genera and families disappeared. However, there was a good chance that a representative life-form remained from each group. Occasionally, there were drastic reductions in populations and the fossil record for a species ceases, only to reappear millions of years later – the so-called Lazarus effect. Although cousins were removed from the battlefield, the family name stood proud while its decimated, remaining species whimpered forward. Although related groups disappeared, it seems that no major phylum was ever eradicated in a mass extinction.

The manner in which these life-forms were lost is important, for it indicates a method of 'testing' nature's designs at the same time as promoting new ones. It is analogous to a series of hypotheses which attempt to explain a set of data. When a new piece of information is found, many unproved hypotheses fall by the wayside. Those which remain are initially fewer but soon generate even more, new theories. In the same way, because no phylum was lost the basic body plan remained. Each mass extinction may have caused the catastrophic loss of a huge proportion of earth's species, but there was always a wide range of surviving life-forms. Every time a species was removed from the scene a greater number arose, to evolve under the new conditions on earth. In biodiversity terms, it was one step backwards but two forwards. Through time, by whatever means, species have increased in number and, at its pinnacle in our recent past, the earth possessed more species at a single instant than at any previous time.

Here, at least one anomaly in the previous discussions arises: for evolution to produce a new species (the process of speciation) this implies that the older life-form is abandoned and, effectively, becomes extinct (unless it continues as a divided population within its own biological island). At the same time, its genes have been passed on to successive generations: the ancestor–descendant chain is unbroken. Speciation can therefore cause a form of pseudo-extinction.

A second cautionary tale involves the rate at which animals and plants become extinct. This is something which occurs all the time, naturally; it is part of the evolutionary process. How, then, is the 'natural', background extinction rate to be separated from a newly imposed rate which leads to what we term a mass extinction? Could it be that, rather than the rate of species loss being reduced, it is the rate of species production which slowed or ceased? Thankfully, aside from causing some lateral thinking, such ideas do not alter the reality of what we understand to have occurred: the numbers of species on earth were reduced in huge numbers.

It is also important to consider why some organisms survived while others did not. How did mammals avoid the effects of the catastrophe which took away the dinosaurs? What factors controlled which groups survived the international extinction lottery?

To make any suppositions here, we must return to concepts developed from observing today's ecosystems. One of the most obvious concerns the food chain. If, as is surmised, plants (as individual specimens or as species) were unable to photosynthesize, their removal would be devastating for any herbivores which relied on them for food. Today, it is estimated that up to forty species might depend on a single plant species, so even a temporary loss of a plant could obliterate a cascading series of life-forms as the food chain is traced upwards. An animal which possessed a varied diet might have more chance of survival than one which had adapted too tightly to a niche.

The reason for selective losses around the world may lie with climatic changes, if extinctions followed the effects of a slow cause. The movement of continents through polar regions would severely affect land-living organisms unable to cope with cold; while India collided with Asia the Antarctic was freezing over and its terrestrial life was lost. As conditions oscillated between warm and cold, with glaciers encroaching over temperate regions, life was forced to migrate or adapt; in many conditions, we surmise, it could do neither. As glaciers retreated, water flooded lands when sea levels rose; what now of sessile, aquatic organisms trapped in deepening water, or those left high and dry as oceans receded once more? Any changes such as these could trigger extinctions, while simultaneously releasing niches to mammals and the forces of evolution.

The earliest mammals were small and shrew-like; as with many first colonizers, they were probably opportunist species, well able to exploit their environment. Could they have been the cause, rather than the result, of the disappearance of the dinosaurs? It has been suggested that the mammals might have developed a taste for dinosaur eggs, challenging their larger rivals for supremacy of the earth by eating their unprotected progeny. By implication, such changes in the balance of species would have been slow and, while it might have caused a decline in dinosaurs, the evidence still points towards an unexpected, sudden catastrophe which impinged on some crucial difference in anatomy or metabolism between what became the survivors and victims.

Mammals are characterized by such attributes as hair, bearing live young, milk secretion, and the possession of warm blood. This last feature is misleading: warm-blooded mammals are those which maintain a constant body temperature, as opposed to the fluctuating, environmentally controlled temperature of cold-blooded animals.

As with many characteristics, being warm-blooded confers both advantages and disadvantages. After a cold night, while reptiles must sunbathe to warm their body chemistry to the point where enzyme action and activity is feasible, mammals can move freely. On the other hand, any animal which maintains a temperature that is warmer than its surroundings will lose heat, which must be replaced. This is the direct reason for the presence of hair and fur (and the use of feathers in birds) as insulation.

Many warm-blooded animals also have complex mechanisms to avoid a fluctuating internal temperature or stresses on the system: sweating in man, panting in dogs, and the use of hibernation and migration. Blood vessels in the mammalian nose are arranged so that they warm air as it enters the lungs and remove heat as it leaves; arteries and veins in the feet of penguins twist around each other so that blood reaching the feet is cooled, heat being transferred to the ascending veins rather than lost to the ice. However, even with these heat-retention mechanisms, the food requirement of warm-blooded animals is much higher than that of cold-blooded ones; a major proportion of the energy derived from food, emanating from the liver as heat, is required to maintain a constant temperature.

This seems to argue that mammals, in a time of food shortages, would die out sooner than dinosaurs, but this ignores aspects of body size. To form comparisons between mammals and dinosaurs, we can study modern warm- and cold-blooded species, making the reasonable assumption that differing food requirements due to an animal's metabolism would be the same then as now. As a generalization, a small warm-blooded animal requires the same amount of food energy as a cold-blooded one ten times its size, due to the need to maintain a constant temperature in the face of continuing heat losses through the skin (because of the relative proportion of the area of skin compared with the animal's body mass, small organisms lose heat faster than large ones). Another way of looking at this is to consider that warm-blooded animals must eat fifty times their body weight a year, while cold-blooded animals only need to eat five times their body weight a year to maintain their life style.

Most dinosaurs of the late Cretaceous were immense herbivores in comparison with the mammals of the time – the shrew-like *Zalambdalestes*, from Mongolia, was only 15 cm from its nose to the tip of its tail, for example. One theory as to why the mammals survived relies on the relative sizes of mammals and dinosaurs: most dinosaurs were more than ten times larger than the mammals, and therefore required more food. In times of food scarcity, small warm-blooded mammals could continue to forage and hunt the available food – which

Dinosaur remains possess more variety than simply fossil skeletons. These *Ovaloolithus* eggs from the Upper Cretaceous offer conclusive proof that dinosaurs were egg-laying. It has been suggested that this characteristic may have helped lead to their downfall, as eggs would have been relatively easy prey for the small mammals which arose at the end of the Cretaceous.

was sufficient for their requirements – while their larger companions starved as they struggled for survival, unable to satisfy their greater demands. Young (and therefore smaller) dinosaurs would, of course, require less food and therefore survived, but as they grew they might have been unable to attain adulthood and therefore breed.

In addition, there is a long-held assumption that the dinosaurs were cold-blooded: some may have developed warm blood. The 'evidence' for this is based on such factors as anatomy, teeth, and similarities to known warm-blooded creatures (dinosaur descendants include our modern, warm-blooded birds, which still have scales on their legs). For example, their posture indicates the continuous use of energy; they could not sprawl like modern lizards, but were able to run erect, with speed.

Further support for this theory comes from the differences in energy requirements of warm- and cold-blooded animals. Because having warm blood requires so much more food, a warm-blooded carnivore requires a larger prey population to hunt than a cold-blooded carnivore does. This, effectively, produces an imbalance: today's terrestrial vertebrate animals consist of only 2 per cent warm-blooded predators but a total of 20 per cent cold-blooded predators (the other 78 per cent are prey). By studying dinosaur teeth, we can deduce their diet and assign a label: carnivore or herbivore, predator or prey. Under 3 per cent of dinosaur biomass turns out to be carnivorous, a proportion linked far more to warm blood than cold.

While this 'warm-blooded dinosaur' conclusion is far from proved, some species of dinosaur certainly do possess features which indicate a specialized mechanism of heat control. This is far from saying that they were truly warm-blooded, but evolution would be strange indeed if it reared such huge, successful animals without there also being a move towards a more efficient means of controlling chemical reactions. Without this, dinosaurs would have been hard-pressed to survive at more than an extremely sluggish pace at the temperature extremes which we now know them to have been capable of operating under. Even many modern, 'cold-blooded' insects can regulate their temperature, especially heavy-bodied flying species. Bees operate and maintain a constant body temperature of 37°C in flight, whatever the ambient temperature, and it is conceivable that the half-metre wingspan, giant dragonfly-like species of the Permian were capable of the same thing. If dinosaurs *were* warm-blooded, their calculated food requirements become higher and they would have been even more likely to succumb to extinction in an era of dwindling food.

In military terms, a first strike is a surprise attack which, hopefully for the aggressor, does not permit retaliation. In true aggression, the invader picks a suitable target: one which does not have an equivalent capability to resist. The first strike weakens the target and takes out some crucial element so that subsequent actions will be successful. There is an analogy here: did a comet provide an unexpected first strike, from which biodiversity could not recover? Was the subsequent action that of volcanic activity, which delivered the *coup de grâce* to the survivors, which were perhaps already weak and unable to respond? They disappeared like wheat stalks sliced during harvest. We still seek answers, but each one gained reveals yet more questions and theories. The solution to our mystery remains encased in a broader enigma.

The extinction lottery clearly favoured some characteristics, and not others. Evolution selects the best of these to suit a niche within a habitat within the environment. Change the environment, and the 'best' characteristics become different. In some instances, the change may not make any difference to a species. Insects may be a case in point: they have diversified into our modern world's most prolific species, and over 80 per cent of those families trapped in ancient, 100-million-year-old amber are still roaming the earth. For whatever reason (their size, capability of flight, energy requirements, independence from water), insects have maintained their capabilities and, in turn, supported the diversification of other organisms such as the plants which they pollinate. They are the true survivors.

A few things are clear: if the species is widespread, it has a good chance of surviving localized changes that will exterminate isolated species. If its gene pool is large enough and its characteristics are many and varied, then it has a good chance of surviving a planetary change. As to why some characteristics are 'better' than others, the question has no meaning without a link to the pressures being exerted: for *this* newly created habitat they are suitable, but for *that* they may not be. As we learn more of the precise conditions in earth's past, and the metabolism of the species involved, we may finally reach a consensus as to why some survived and others did not.

As to modern extinctions, it is easier to point to the cause: man's influence on the environment. Pollution and habitat destruction – draining wetlands, building dams, logging – are just two of the more obvious factors which affect organisms locally and globally, a theme which is returned to in the final chapter. What is less often realized are the insidious effects of hunting and the introduction of competing species.

In 1894 an anonymous description was given of the Passenger Pigeon of North America, part of which quoted the writings of the naturalist, Audubon, from some 80 years earlier. By the time the book was published, the pigeon was already doomed:

> *In these almost solid masses they darted forward in undulating and angular lines, descended and swept close over the earth with inconceivable velocity, mounted perpendicularly, so as to resemble a vast column, and when high were seen wheeling and twisting within their continued lines, which then resembled the coils of a gigantic serpent.*
>
> *The air was literally filled with pigeons, the light of noon-day was obscured as by an eclipse, and the continual buzz of wings seemed to lull the senses.*

The splendid sight was not one which would last: pigeons were such a bountiful source of food, they were hunted without mercy. It was a spectacle which is hard to comprehend today, so vast were the numbers. Arriving at one roost shortly before sunset, Audubon noted that:

> *Few pigeons were then to be seen; but a great number of persons with horses and waggons, guns and ammunition, had already established encampment on the borders. Two farmers from the vicinity of Russelsville, distant more than a hundred miles, had driven upwards of three hundred hogs to be fattened on the pigeons that were to be slaughtered. Here and there the people employed in plucking and salting what had already been procured were seen sitting in the midst of piles of these birds. . . .*
>
> *Suddenly there burst forth a general cry of 'Here they come!' The noise which they made, though yet distant, reminded me of a hard gale*

at sea passing through the rigging of a close-reefed vessel. As the birds arrived and passed over me, I felt a current of air that surprised me. Thousands were soon knocked down by the pole men. . . . The pigeons, arriving in thousands, alighted everywhere, one above another, until solid masses as large as hogsheads were formed on the branches all around. Here and there the perches gave way under their weight, with a crash, and falling to the ground, destroyed hundreds of the birds beneath, forcing down the dense groups with which every stick was loaded. It was a scene of uproar and confusion. I found it quite useless to speak, or even to shout to those persons who were nearest to me. Even the reports of the guns were seldom heard, and I was aware of the firing only by seeing the shooters reload.

Despite numbering in millions the Passenger Pigeon (*Ectopistes migratorius*) was extinct in the wild by the end of the nineteenth century, with the last zoo specimen dying in 1914. Other species went the same way: the Carolina Parakeet (*Conuropsis carolinensis*) in the USA and the New Zealand moa family of flightless birds, lost to Maori hunting. Hawaiian treecreeper species, hunted for plumage and deprived of their habitat, have suffered to the extent of nearly half their species disappearing this century. Since humans arrived in Australia, nearly one-third of all mammalian species have become extinct.

The classic example of hunting to extinction is that of the dodo, in reality a family of three

It is virtually impossible for man to avoid environmental damage – all that can be done is to minimize the worst effects. Hydroelectricity is billed as a clean source of power, as at Kariba Dam on the Zambia–Zimbabwe border, but when the land is flooded habitats are destroyed. Across the lake, which was completed in 1960, the stark remains of trees still jut from the water. Perhaps the river god, Nyaminyami, is also distressed: the native Batonka people were driven away, and Operation Noah was mounted to save some 5,000 animals from drowning. This, however, resulted in overcrowding on the southern bank and intense competition between animals until the populations stabilized once more.

There will be no easy recovery after the remaining forests of Madagascar are felled. Some trees have been dated to 800 years old, and cannot be replaced by replanting.

large, flightless pigeons. The species separately inhabited three islands off Madagascar, speciating as they adapted to local conditions. Discovered by the Dutch and Portuguese in 1598, these turkey-like birds were extinct on Mauritius by 1681, on Réunion by about 1750, and on Rodriguez by 1800, prey not only to hunters but also to introduced, competing species, such as dogs, pigs and rats, which scavenged their egg-laden nests.

There are many examples where a new species brought into an island community can have a devastating effect. Of particular note is the introduction of herbivores such as goats and rabbits, which destroy the plant community which underpins any food chain. In 1870 rabbits were introduced to Stewart Island, off New Zealand, where rats and cats had been set loose by sealers. The rats and cats fed on ground-nesting birds, and the rabbits ate the tussock grassland which reduced the area of breeding habitat. In an attempt to control the damage, the myxoma virus was put to use in 1968 with, instead of mosquitoes, the introduction of the rabbit flea as a vector. Rabbit numbers have since dropped from over 150,000 to less than 15,000, and tussock grass growth has improved. However, it was too late for two species, a parakeet and rail, as they had already been driven to extinction by man's introduced pressures. A similar tale arises from feral cats, introduced in 1949 to Marion Island in the South Atlantic. By 1977 the number of cats approached 3,500, and they were controlled with the introduction of a feline virus. Those 600-plus which survived, exhibiting immunity, were shot or poisoned, but not before the population of burrowing petrels had been decimated: in 1975 alone, 450,000 birds were killed.

Problems from introduced species are not restricted to opportunist scavengers. A species of flatworm which coils itself around earthworms, strangling them to death, has reached Europe's ecosystems from its native home of New Zealand. *Artioposthia triangulata* has become established in Ireland and Scotland since the 1960s, and is spreading southwards. At risk is the valuable work that earthworms accomplish in introducing organic matter and turning over topsoil and, of course, providing food for species such as owls, badgers and moles. The flatworm has no natural predators and, although it is not pushing earthworms to extinction, its effects are all too clear. Similarly, the Cane Toad, *Bufo marinus*, was introduced from Hawaii to Australia in 1935 to control a plague of beetles. The toad failed to tackle the problem – but proved capable of eating any native animal which would fit into its mouth. Bearing poisonous glands, the Cane Toad is an ecosystem nightmare which is spreading from Queensland through the continent's Northern Territory.

The gene pool is inevitably damaged when competition and hunting occurs, even when their effects stop short of total elimination. Losses due to introduced species stand at about five times those attributed to hunting and collecting, but devastating effects can be produced locally. When, in 1810, Captain Frederick Hasselborough landed on the Australian island of Macquarie, he was able to take 20,000 fur seals in the first 18 months, eradicating the species by 1822. Elephant seals followed, then king penguins, which were boiled 2,000 at a time in man's greed for oil. Whaling's history is long and bloody, yet these magnificent mammals are still persecuted. There is little to be proud of in the continuing history of our approach to life.

To return to the concept of a species' expansion into an empty niche, it is interesting to note the extent of the age of dinosaurs, a spectacular success story rather than a sad failure. The dinosaurs rose to power when the mass extinction of the Permian–Triassic 250 million years ago created empty niches for them to enter. When in turn the majority of the dinosaurs disappeared only 65 million years in the past, it was their niches which became empty. Perhaps it was, ultimately, an act of God in the form of an unpredictable comet which cleared the way for the explosive, adaptive radiation of the mammals. Without this, evolution could not have created man. Yet, while without sufficient humility (and certainly falsely) we sometimes consider ourselves the top of the evolutionary tree, it is worth remembering that the dinosaurs were dominant for over 150 million years, while our ancestors have walked erect for merely two million years.

The small size of the early, warm-blooded, shrew-like opportunist mammals may have been the crucial factor which permitted them to inherit the earth from its previous custodians, the dinosaurs. However, the age of those terrible reptiles pales into insignificance when the true survivors are considered: insects in fossil form have been dated to 390 million years old. When the evolutionary goalposts were moved for other species, it seems that the insects continued along their own road.

To the phrase 'The meek will inherit the earth' must be added 'Only when the insects have finished with it'.

CHAPTER EIGHT

GARDENS OF PLENTY

Life, life, everywhere life!

Arabella B. Buckley, *Winners in Life's Race*, 1892

Produce of more than one country.

Supermarket label, 1995

THE EUROPEAN COURTS of medieval times were well fed. The Romans had already brought fruit trees, cabbages and vines; gentry ate herb fritters soaked in honey, and purées boiled in almond milk. Saffron, from crocus stigmas, was used to flavour food and tint bathwater (in Britain, the principal centre for growing crocuses was Saffron Walden). Nobility used ginger, cloves and cinnamon to spice their meaty dishes. There was a demand for more, and greater, variety – and lower costs. However, the lucrative spice trade with the East was dominated by Arab and Venetian middlemen. Enough was enough: Europe turned to the oceans to find ways past their monopoly. England, Spain, Portugal, Italy and Holland became major sea powers, driven by a taste for the exotic; where one spice was found, there must be others. In their time Cabot, Columbus and Magellan hunted for new routes to sources of nutmeg, pepper and cloves, and the sciences of astronomy, timekeeping and magnetism were developed as part of the necessities of navigation, driven by the search. But, as far as spices were concerned, there was little success: the ventures ended in failure, though the dream went on.

The Renaissance did, however, bring about the discovery of the New World: the Americas and the East. In place of spices, there was the potato, which was grown by the Incas at high altitudes where cereals could not survive. Sir Walter Raleigh is credited with bringing back the starch-rich food, but the potato did not initially attain popularity at court as, at a banquet held by Elizabeth I, the cook only served the leaves and stem. As the potato is related to Deadly Nightshade, all green parts (including sun-exposed tubers) are poisonous and the 'new food' produced little more than sickness: it was relegated to feed cattle and the poor.

The French brought back a bean which proved exceptionally popular and was named after them, while the tomato, known as the 'love apple', was initially (outside Italy) grown only as an ornamental plant. Green peppers, a relative of the potato and tomato, were also welcomed from South America, together with avocado, peanuts and squashes. Sugar cane, a native species of the Indian sub-continent, was spread to China and Central and South America, reaching south-eastern Africa in the 1600s and Australia in the 1800s. Oranges (a Sanskrit name, given to the fruit by Arabs) were first taken from South-East Asia by Moors and used in medicines in Spain, spreading through the New World in the sixteenth century. When supermarket labels indicate 'Produce of more than one country', there is more truth involved than even the advertising copywriter imagines.

With its knowledge of the plants which grew in the Indies, in 1600 the British East India Company was chartered by Queen Elizabeth I with the intention of breaking into the Dutch-dominated spice trade operating throughout Indonesia. It went through setbacks and strengths, even being given Bombay by Charles II and the right to mint coins and go to war against any non-Christian power. Trade in silks, cotton, spices and saltpetre (used in the manufacture of gunpowder) was excellent, and the great commercial machine turned a tidy profit as it spread further afield to China and other parts of the East.

Other countries did not stand still: the Dutch fought for their established Indonesian trade, and repelled the British while displacing the Portuguese in Asia. The Dutch East India Company developed a monopoly with Japan and the Dutch colonies of the African continent. The French also possessed an East India Company, operating between India, Mauritius and Réunion. The powers of commerce were great: it was the hunger for exotic spices which caused the Europeans to become major sea powers. All this drive and effort – for plants.

As part of normal operations, company employees were encouraged to search out new species which might prove economically valuable. The British East India Company shipped tea from the East and sugar from the West to satisfy its home market; specimens and drawings were produced, many of which are now central to museum collections. From China came 2,000 commissioned illustrations by

Poppies are used as sources of seeds for cooking, and drugs are made from the latex derived from the sap. The Latin name for the opium variety, *Papaver somniferum*, has links with Roman mythology: Somnus was the god of sleep and a twin to the god of death. His abode, a cave near the underworld river of Lethe, was rich in sleep-inducing poppies, and Somnus is often depicted as carrying a poppy stem. His son was Morpheus, hence the name of one poppy-derived drug, morphine.

Chinese artists, and expeditionary voyages added to the spoils. Bananas spread from Malaysia to Africa and the Americas, while tobacco was exported from South America. The same story can be told of cocoa and pineapples (originally from South America), and coffee (Africa). Vanilla, from Mexican orchid seeds, was brought back by the Spanish. The pretty red poppy, a native of the Mediterranean, was spread across the world, cultivated and had become the commercially grown opium variety by the late seventh century. In the 1770s the East India Company, following in the footsteps of the Portuguese in selling opium to China, grew the plant and, by circuitous routes, illegally entered into trade. Opium came to Britain under its regime, with a host of other plants; London and Amsterdam were the drug capitals of the world. In the 1800s, to avoid taxes, aloes were smuggled inside monkey skins, sarsaparilla in South American cowhide. Nutmeg was so valuable that imitation nuts were carved out of wood, dipped in nutmeg oil to give them scent, and sold as the genuine item. And, in India, there was the basis of wealth: suitable lands for plantations and a cheap workforce, ripe for exploitation. All that was required were new plants and seeds.

Where nature is bountiful, man has always found the means to exploit it. At its most basic, there are food, minerals, shelter: here is a sustenance crop or exotic, seasonal fruit; there, the wood to burn as fuel for smelting iron. Wood is pulped for cardboard and paper (we've come a long way from Egyptian papyrus). Plant fibres clothe us and, as medicines, cure us from ills, while industry runs on seed oils and lubricants. Plants even pervade our language: corny jokes, nutty folk, playing gooseberry, he's got fibre.

Man, the hunter-gatherer, has today vastly extended the original concept of barter. When organizations such as the East India companies began their operation, they were doing no more than extending systems of trade which had existed since the earliest days of prehistory. In doing so, for reasons of commerce or simple curiosity, they spread plants and animals around the globe from their original habitats to new areas. One facet of these activities is certainly interesting: plants ensure their own existence using varied techniques of seed dispersal, which is accomplished extremely efficiently under man's cultivating hand – so which species is truly using which?

As an example of this cultivation-dispersal, in 1787 Captain William Bligh was commissioned to sail to the Pacific island of Tahiti at the suggestion of Joseph Banks, the naturalist who accompanied Captain Cook on his voyage of 1768. Banks studied the requirements of the day, and believed that the breadfruit tree would form a suitable crop to feed the sugar plantation slaves of the West Indies. In 1789 the trees, which grow to 18 m or more, were placed in the hold and Bligh set sail for Jamaica; he never arrived for, as has been retold in books and films, his crew on HMS *Bounty* mutinied under Bligh's tyrannical rule and set him adrift in an open boat. Bligh completed his mission during another voyage in 1793. From an oval, seedless fruit crop of Polynesia, breadfruit has spread throughout the tropics.

The degree to which we rely upon just a few species of plants for food is astounding. Think of the many fruits and vegetables on sale in greengrocers: apples, pears, beans, peas – and mangoes, kiwi fruit, guava and okra. Now consider more basic foodstuffs: bread, rice, potatoes. It is perhaps surprising, given the vast numbers of plants on earth – over 10,000 species are known to be edible – that we rely on only about 150 for cultivation. Of these, 20 species provide 90 per cent of the world's calories. The cereals alone – rice, wheat and maize (often also called corn, a term which is

Opposite: The street markets and greengrocers of today sell both imported and local produce. The ancestral plants which supply these fruits and vegetables, however, probably originated far from today's crop-dominated fields. This trader in Iráklion, Crete, may be an exception for at least a few species, as the Mediterranean was historically a centre for a number of man's crops.

1000
300

commonly applied to any cereal crop) – supply over half the planet's food requirements and cover three-quarters of its farmlands. Add two more, potatoes and barley, and you have eight species supplying 75 per cent of all foodstuffs.

By way of comparison, for meat we rely on only a few animals, ranging from pigs and cattle in first place followed by poultry, sheep, goats and water buffalo. With the addition of camels, horses and mules, virtually all food animals of note, in terms of world production, are now listed. 'Don't count your chickens . . .' may be prophetic: we rear more than double the number of poultry animals than there are humans on earth, yet the total of all animal husbandry supplies well under 1 per cent of the planet's food energy.

Cereals are named after a Roman spring festival, the Cerealia, held in honour of the goddess of agriculture, Ceres. Wheat and barley were cultivated in the Middle East at least 10,000 years ago, and the Romans conquered Egypt to gain access to its fertile wheat fields. Cereals are types of grass, of which there are over 8,000 species, and to the 'big three' must be added sugar cane, bamboo, oats and rye. These last two grains were originally classified as weeds, being spread with valuable wheat and barley grain, it only later being realized that they grow better than other cereals in colder, moister, northern climates, where they have become an important crop.

Rice, meanwhile, forms the principal food for about half the world's population, the commercial variety originating in the Indian sub-continent rather than China. Rice requires a lot of water for growth, hence the flooded paddy fields of Asia, but the water's main function is to drown competing weeds. Rice can therefore, with care, be grown in drier conditions, and endeavours are being made to increase its growing range.

Such experiments are simply the end of a long line of attempts to domesticate crops by artificial selection, and few of our food plants remain unaffected. The humble carrot is a native of Afghanistan, being spread through Europe and the world as naturally coloured red, purple or black varieties. In the sixteenth century a pale-coloured strain, lacking the anthocyanin pigment, was used in soups as it added flavour without colour, while the modern, rich-coloured carrot strain was developed in seventeenth-century Holland. When North America was colonized, the carrot was taken as a food plant. It escaped from garden plots to become a wild flower, Queen Anne's Lace, an example not only of the spread of plants but also the production of new strains. In the same manner, the biblical symbol of temptation and the scientific emblem of inspiration, *à la* Newton, exists as upwards of 7,000 varieties of apple, most of them developed in the eighteenth and nineteenth centuries.

On the face of it, artificial selection (such as the intentional production of a carotene-rich carrot) is doing no more than helping increase variation. Surely, a larger fruit or stronger growth is beneficial for the plant as well as man. In fact, this is rarely the case. As each generation matches what man desires a little more closely, most plants become more and more dependent upon being sown and tended. Frequently, in attempts to increase productivity, plants lose their ability to grow naturally in the wild. In particular, their reproductive mechanisms may be damaged in the age-old attempt to domesticate, rather than simply cultivate.

Wheat no longer needs a period of dormancy before it will grow. Dormancy is a requirement of many plants, in which the seed must go through a period of cold before it will germinate. This prevents the plant from beginning growth at the start of a mild winter, only to be destroyed by frost before the spring. Other domesticated plants lose protective features, such as poisons, or even the ability to produce seeds. Bananas are one example of this – yet the only way to produce new (seedless) trees is to grow them from seed, using the ancestral species or searching through fruits for an occasional reversion. Cultivated potatoes have, behind them, a vast genetic resource, but are sterile: 'seed potatoes' produce no more than clones of their parents, and ancestral genes are required to breed new hybrids or confer disease resistance.

The Western world's kernels of sweetcorn comprise a highly modified genetic product. The five major varieties now grown are dependent on threshing to remove the seeds; there is no inherent dispersal mechanism and, indeed, it was the aim of the artificial selection process to produce a plant which *could* be efficiently harvested. To match the requirements of modern cultivation, wheat must grow to a uniform height so that it can be easily reaped by combine harvesters; the plant is made to suit the machinery, and not vice versa. Wild varieties of wheat readily become detached from the stalk when ripe, and possess long spikes on the grain-rich ear to help them to catch on soil and begin germination. The ears were too brittle for mechanization, and the spikes interfered with reaping – so both features were bred out of the plant by turning the wild varieties into polyploids with six sets of chromosomes.

In 1913 Leonard Bastin commented on the success of the plant breeder who 'has given to the world a carrot which produces a large and shapely root' and 'a spineless cactus' for cattle food in desert conditions, but bemoaned the increasing loss of scent from new varieties of rose. With each modification, plants are removed one step further from their genetic origins. With genetic damage and the ability to survive and evolve in the natural world severely impaired, our food plants are indeed dependent upon their masters.

Another factor is raised by this dependence on just a few 'super-species'. The world's few crop species are effectively grown as monocultures; every attempt is made to eradicate any competing species such as poppies or other 'weeds' in cornfields. More than that, as the world's populations continue to rise, greater food productivity is required and the monocultures become even more of a necessity. The production of 'alternative' forms of food is slowly being strangled, to leave other species with the status of cash crops. As the traditional varieties of food diminish in stature, man is cramming more eggs into fewer baskets.

Gila Cliff Dwellings in New Mexico were built some time after AD 1000, but the site was home to Mogollón Indians before the time of Christ. When the structures were examined, the small room to the right was found to contain dried remains of maize cobs, a clear sign that cliff dwellers were farming the area using domesticated crops. The dwellings seem to have been abandoned at the end of the thirteenth century.

'I beheld with sorrow one wild waste of putrefying vegetation. In many places, the wretched people were seated on the fences of their decaying gardens, wringing their hands and bewailing bitterly the destruction that had left them foodless.' It was 1846 when Father

Matthew, an Irish priest, wrote of the total loss of the potato crop for the second year running. From every indication of a bumper harvest, within a week the plants were dead; the trend continued for year after successive year. This was the time of the great famine, which reduced Ireland's population – by reason of starvation, disease and emigration (it was as a direct result of the tragedy that many Irish people left for New York) – from over 8 million to nearer 6 million.

The potatoes had been attacked by a parasitic fungus, *Phytophthora infestans* producing potato blight; it was the study of the blight which led to the first confirmation that microbes cause disease, rather than being a symptom of decay. As it grows inside the leaves, the fungus spreads and erupts through pores in their surface. In turn, spores are released which infect other plants. Once present, the fungus remains in the soil, ready and able to infect subsequent attempts to grow potatoes. Where one field is lost there is economic ruin for a farmer; where the staple diet of a country is destroyed, there is a catastrophe.

Coffee forms a valuable cash crop, yet the plantations remain open to disease: over 99 per cent of coffee beans originate from only two of the many species of the *Coffea* genus. The distinctive tastes which the beans yield depend not only on the species, but also its growing conditions (the best come from volcanic soils between 900 m and 1,800 m in altitude) and the method of roasting.

The difficulty lies with our penchant for monocultures. It is similar to human disease: pack too many people in close proximity to each other and the likelihood of catching the common cold is greatly increased. Grow a single crop over a widespread area, alter its genes so that it is dependent upon certain, narrow conditions being maintained, and there is a recipe for disaster. Bend a species to our needs, and its genes become eroded and weakened.

Surely, in the last 150 years, we have learned how to cope with such disastrous crop failures as experienced by the Irish. Today, we rely upon an annual world production of more than 300 million tonnes of potatoes; if the number is too enormous to make any sense, it's the equivalent weight of 60 million elephants, or enough potatoes to circle the globe 500 times. We cope with the plant's weaknesses using knowledge of the diseases which affect it (potato blight only occurs when the weather is moist and warm, so predicted weather patterns enable farmers to spray crops at the most likely times of infectious growth), rather than correcting those problems or using more varieties.

However, we haven't learned from the lessons of monoculture, and even rely on planting clones of the parental potato as a crop. In the 1860s Europe's wine industry came to a staggering halt when diseases attacked the vines. For 20 years, from 1870 to 1890, Ceylon's coffee cash crop was destroyed by another fungus: coffee rust. This century has seen millions of Bengali people die when the rice crop failed in 1942, while the oat crops of the USA were crippled by disease in 1946. Each time, as with a 1950s' wheat stem rust in the US, fungus was the essential, devastating factor. Each time, the epidemic was predictable.

The organisms which cause disease are as much a life-form as their hosts; they mutate and evolve under the same rules of evolution. If a fungus, bacteria or virus penetrates a wild plant's defence system, bypassing its protective coat and poisonous chemicals, it rapidly reproduces and attempts to infect another plant. There is one, all-important factor which acts in the plant's best interests: it has evolved alongside its attacker, and its genes contain the characteristics required to repulse the invader – if not in every plant, they are present in some. In the wild, with populations spread over large areas and with variety between them, even if some of the species succumb, a rich genetic history forms the ultimate defence. The survivors of any attack hold a valuable gene combination which passes to their offspring.

Until man, and monocultures. Here, the gene combinations are fewer, while the parasites are as virulent as ever and, with a shorter reproductive cycle, can mutate into new strains with alarming speed. Resistance is low, and there is no recovery from an attack of epidemic proportions. The plants, with a fading gene pool, cannot survive even with the aid of pesticides. The average lifetime of commercial crop varieties is brief, as little as between five and fifteen years in the case of wheat, before a new strain must be developed in the face of disease. With inbreeding comes a decrepit stock, and there are no known instances of a crop strain being able to compete in the wild with its genetic ancestors.

In 1970 a fungus attacked the US maize crop, which represented 40 per cent of the world's harvest. Over 80 per cent was destroyed and it appeared as if maize, its genes having been heavily manipulated since the 1920s, would cease to be an economic crop when losses due to fungal attack continued. Then, in 1978 a 'new' variety of maize (*Zea diploperennis*, referred to as the Teosinte strain) was located in the Sierra de Manantlan of Mexico by an undergraduate student cataloguing the country's flora. In fact, the Teosinte variety (in appearance its weedy ears of grain are like those of grass, of which – of course – it is a species) proved to be one of the US maize crop's ancestral strains, carrying the essential genes to confer fungal resistance.

Genetic erosion and biodiversity have suffered drastically as food sources become standardized. Of over 100 varieties of cattle, only 30 breeds are maintained. Chickens are bred and inbred to produce more eggs and develop added

Monocultures have the advantage of a higher yield, but generally suffer from a reduced resistance to disease and a decreased gene pool. Only 20 species of plants supply 90 per cent of the world's food.

flesh for the table, while the traditional, uncommercial varieties slip into obscurity. When new genes were required to bolster the inbred stock, only a few, now valuable, fast-growing subspecies were available. So it was with the Teosinte maize: only some 2,000 plants were found in a small area of Mexican farmland, and its habitat was decreasing. If the plant had been eradicated, its genes gone forever, US crops would not now be protected from over half the known maize diseases.

Every major crop is in the same situation. Despite Andean farmers recognizing some 5,000 varieties of potato, we rely upon just six strains of one species for 80 per cent of our production. In Canada, half the wheat production is dependent on one strain, while in the eastern Mediterranean the number of varieties of wheat being grown has dropped from around eighty to below ten. Yellow dwarf disease affects barley; our crop protection is aided by the introduction of a gene discovered in a Ethiopian strain. Our developed varieties of cauliflower, cabbage, broccoli, kale and Brussels sprouts (first grown in quantity near the Belgian city of Brussels in the sixteenth century) may in the future rely on the infusion of genes from plants such as the ancestral wild cabbage, *Brassica oleracea*; unless, that is, another plant has been discovered in the meantime.

The principle of combining genes from different strains is a common one. The sunflower seed, used as a source of oil, is grown by plants which depend on gene input from a wild variety. The soya bean was first used commercially for the manufacture of paint, varnish and soap, as it contains about 20 per cent oil. Later, its 40 per cent protein content (the highest of any bean or pea, and five times that of maize; soya beans are similar to meat in protein content) earned it today's position as the largest cash crop of the USA. Yet, the domesticated soya plant is a combination of the 'best' characteristics of six Asian strains.

Dependence on a very small proportion of the gene pool is also very common, due to the use of only a few collected specimens for cultivation. Many of these plants, such as rubber, are commercially valuable.

Columbus is said to have first observed Haitian children playing with balls of elastic-like material, produced from an extract of

Seeds from sunflowers, such as these growing in the Loire Valley in France, are used as poultry food and a source of oil for manufacturing margarine.

trees. The product, rubber, was given its name by Joseph Priestley as it could remove pencil marks. Rubber is obtained by cutting the bark of a tree (several species are suitable), causing its sap to seep out for collection; its American Indian name, *cachuchu*, is therefore quite descriptive: the 'wood that weeps'. Initially no more than an interesting novelty, in 1823 Charles Macintosh used the resin to waterproof clothing (hence the mackintosh garment). In 1839 the process of vulcanization, which adds sulphur to natural rubber to make it hard and resilient, was invented; another Charles was linked to the rubberized automobile tyre: Goodyear.

The rubber tree, *Hevea brasiliensis*, was first grown commercially in Asia by the British as its supply from Brazil was not dependable. When Brazilian legal restrictions caused further difficulties, the British government hired one Henry Wickham to acquire some seeds from the rainforests of South America. These few extracts from the gene pool were subsequently grown in England in 1876 before the seedlings were transplanted to Ceylon (now Sri Lanka), Malaysia and Singapore. Ninety per cent of the world's production of rubber comes from these countries, all of it from the descendant plants of those seeds. A similar story can be given for the South American Cinchona tree, also transplanted to the West Indies. Cinchona bark yields an alkaloid drug, quinine, used to treat malaria and heart disease and to stem pain. Likewise, Brazilian coffee plantations can trace their lineage back to a few original trees from a single plantation.

The crux of these arguments lies with our overdependence on just a few species, coupled with an essential need to maintain the gene pool as a resource to forestall future disasters. It should also be evident that the majority of the species we rely upon originate from the tropics and, in particular, from the rainforests. However, we can only reap the benefits from species that we can first find, catalogue and test.

In 1962 a search was conducted in the Peruvian Andes for wild potato varieties by two men, Hugh Iltis and Don Ugent. They found a 'ratty-looking' tomato from which they collected a few seeds for posterity, sending them to a colleague, Charles Rick. By 1976 the plant, the 832nd in the collection, was described as the eighth known tomato species. Over the next few years its genes were used in hybrids and crosses, until new strains with larger fruit, a richer colour and a 40 per cent increase in solid weight were available. The added weight was in the form of sugars: biodiversity had provided the world with a sweet success that, today, even aids in a resistance to the debilitating tomato disease of Fusarium wilt. In commercial terms, the genes are worth over $10 million a year to the tomato industry, for a collection cost of $21 per sample.

The use of plants and animals for medicine is nothing new. For centuries there have been herbal cures and witches' brews, myth mixed with superstition. We no longer believe, as did the people of the 1500s, in the powers of amphibians. If you were unlucky enough to fall and knock your nose against a stone, the bleeding would be instantly stopped if only you possessed a toad which had been pierced with wood and dried in the shade or smoke; it seemed 'that horror and fear constrained the blood to run into its proper place, for fear of a beast so contrary to nature'. Frogs and toads formed crucial ingredients for Shakespeare's witches, plus such powerful substances as a newt's eye and slivers of yew: 'Cool it with a baboon's blood, then the charm is firm and good.' However the ancients felt about these much maligned animals, today we use steroids secreted by the poison arrow frogs of the Amazon in medical treatments, in particular to develop antibiotics and anaesthetics, scorpion venom to build new protein drugs against nerve diseases, leeches for heart treatment, and the funnel web spider's bite to treat stroke victims.

The principle is a common one: where nature produces a poison, it has a specific target. The tissue involved may, if it is diseased, respond to medically controlled, low concentrations of the toxin. The poppy is not only

Poisons secreted from the skin of some frogs, such as this specimen of a Blue Poison Dart Frog, *Dendrobates azurens*, have traditionally been used by South American natives to coat the tips of darts and arrows. Today, these steroids are valued as ingredients in medicines.

capable of offering us opium, but also morphine – from which codeine is extracted as a weaker painkiller. Heroin was discovered in 1898 by modifying morphine, and was freely used in cough medicines until almost the end of the First World War. Steroid drugs were developed from diosgenin found in yams and beans (seeds of the Fenugreek herb, *Trigonella foenum-graecum*, also supply diosgenin for steroid manufacture). The bark of plants is often rich in toxins and exotic chemicals, as it forms the plant's first line of defence against attack by other organisms. So it is that the Pacific Yew yields the drug taxol, used since 1992 for cancer treatment (the poisonous nature of yew has long been known; required for its resilient wood, yew was kept from cattle by growing it behind churchyard walls), and aspirin was first produced from an extract of willow bark, *Salix alba*, before it was found that the Meadowsweet herb, *Filipendula ulmaria*, was a better source. Curare, used on poison arrows, is distilled from South American tree bark; in small quantities it is used as a muscle relaxant during surgery. Atropine, from Deadly Nightshade, dilates the pupils of the eye in small doses; ladies once used atropine to make themselves more attractive, hence the Latin name of *belladonna*.

So, the search is on for the genes of the wild. In the 1930s and '40s the thrust of pharmaceutical research was towards plant-based chemicals, and the switch was to synthetics and microbes (for example, in the production of penicillin) until it became evident that science was failing in its attempts to formulate drugs to combat modern diseases such as AIDS. Countless hours can pass in laboratories trying to synthesize new organic compounds, and then there is no indication what use they might hold. It is, simply, far cheaper and easier to collect existing molecules, constructed under the control of a genetic blueprint, and to extract or synthesize what is discovered. The speed of change is fast: at the close of the 1980s no US-based pharmaceutical company had an active plant research policy. Now, the tables are reversed and many countries, including China and Germany, screen for bioactive plant compounds. To support the programmes, some bioprospectors are stripping animals and plants from land and sea on the off chance that a viable product exists.

Steroid compounds such as cortisone and the drugs used in contraceptive pills were first isolated from glands of cattle, but the cost of production was too high to be realistically prepared. A deliberate search for plant substitutes was initiated; it proved successful with Japanese yams, but quantities remained low. When three Mexican species turned out to produce huge tubers, and therefore could supply large volumes of contraceptive chemicals, they were grown commercially for the purpose. Around half the steroid drugs produced are still made directly from plants, rather than being synthesized in a laboratory. For men, of course, plant-based contraception still relies upon the rubber tree, though cotton seeds may yet hold the key to a male oral contraceptive.

Some unexpected discoveries have been made in recent years; armadillos, for example, can contract leprosy naturally and the animal is used in the preparation of human vaccines. Sharks, hunted for sport or destroyed as the ocean's unwanted and unloved killers, turn out to be useful in liver and cancer research. If you suffer from arthritis, the genus *Oenothera* (which includes the Evening Primrose) supplies an oil which may prove helpful in alleviating pain; tests are proceeding.

Pharmaceutical uses are perhaps one of the most lucrative utilities of life; around three-quarters of all pure drugs on sale can trace their lineage back to plants used in traditional medicine. However, there are other fields of endeavour which are and will prove just as beneficial. For the unexpected, turn beyond the use of microbes.

Copper mine workings dating to before the time of Christ are today leaching metals into the soil. However, rather than the weather being involved in this pollution, in 1947 it was discovered that a bacterium – *Thiobacillus ferrooxidans* – was responsible. It exists by reducing sulphur compounds based on metals such as copper, gold, lead or zinc to the pure element, deriving energy from the breakdown process. In the 1970s Canadian uranium was commercially extracted by bacteria rather than more normal processes, with gold in South Africa and copper in the USA soon following suit. Using bacteria to wrest metals from the soil proved a far cheaper process than smelting.

Are your fruit trees plagued by greenfly? Follow the recipe: ignore the pesticide sprays, call a supplier for a litre of predatory bugs, spread liberally throughout the area, and sit back as the feast begins. Elsewhere, strips of weeds planted near wheat fields attract hover-flies, which lay eggs and produce aphid-eating larvae. As a pollution control, maggots are being used to clean up dumped animal carcasses. Neither is the use of a biological pest control a new idea. In the 1880s the citrus crop

Aphid pests are increasingly being treated by using an influx of natural predators, rather than more costly spraying with pesticides which has to be repeated at regular intervals.

in California was threatened by the Cottony Cushion Scale-insect, which was brought under control by introducing its predator, the Vedalia Beetle. In Australia, where a burgeoning house mouse population was becoming a problem (in cycles, the species can spring from a few individuals per hectare to over 5,000, destroying crops in the process), the latest tactical weapon was recently brought in: the parasitic nematode *Capillaria hepatica*. Nematode worms are also used in gardening; purchase a pack and see the wonderful results in your pest-free roses next season.

Of particular interest was a 1970s attempt to control plant-hoppers on rice crops, which dramatically increased in numbers when Indonesia began to produce two harvests of rice a year instead of one. Huge subsidies enabled pesticide spraying eight times a year, but the problem increased. Then it was found that the spraying was killing the plant-hopper's predators; the spraying was dropped to 10 per cent of its original level, and conservation encouraged spiders. Yields of rice increased; more rice and less pesticides meant more money.

The search for new chemicals and the genes which produce them raises questions. Some plants are widespread and have been for millennia, while others are taken from relatively small island habitats to be carried to plantations at the whim of commercial concerns. Are such species the property of the world, the country where they are found, or of the local people? The rape of Brazilian forests to stealthily remove coffee seeds was one of a different age; is it now ethical to use the gene pool of another country to bolster a failing, foreign industry or create a new, vigorous crop? Where does an ethical approach place the division between lucrative trade and the rights of an indigenous population?

One of the most striking examples of the balance to be struck is represented by the Rosy Periwinkle, from the island of Madagascar. Often grown as a pot plant, *Catharanthus roseus* has become famous as a symbol of biodiversity. A member of the poisonous dogbane family, it contains alkaloid compounds; an overdose is lethal, although a controlled quantity can be beneficial. First isolated in 1958, extracts of the periwinkle (vincristine, vinblastine) are used to treat leukaemia, Hodgkin's disease and other cancers to the extent that patients now have an 80 per cent chance of survival compared with less than 20 per cent only thirty years ago. The cost? The Rosy Periwinkle yields only 100 g of alkaloids, worth over $20,000, from 53 tonnes of leaves.

Obviously, the native home of the periwinkle should be conserved at all costs, for the benefit of mankind. In fact, the periwinkle's natural habitat is threatened and increasingly turned over to subsistence farming. Surely it is up to the Madagascan government to ensure it is conserved – but why should it? While the Western world cures its ills using an African gene, the peasant farmers who work Madagascan soil share as much in the proceeds of the drug as does the rest of their country: not a cent is returned or offered. The gene rustlers have arrived, plundered, and departed.

If a country holds the rights on behalf of the world, what should be done when it squanders its resources? While foxgloves, which yield the heart treatment drug digitalis, are conserved in natural German forests to preserve a source of the flower, other organisms fade into obscurity even when the value of their genes is recognized. The Manatee, a docile herbivorous mammal, is found in the waterways of Florida. It dines on water hyacinth, a fast-growing intruder which chokes Florida's creeks, with the Manatee's genetic value residing in its slow-clotting blood, pointing the way towards research into haemophilia. The Manatee now numbers well under 1,000 individuals, living on the doorstep of the nation which boasts the greatest investment in biotechnological research.

Science is certainly not squeaky clean. As areas of the world are searched for genes, given the low quantities of useful chemicals found in the Rosy Periwinkle (only 0.00025 per cent of the dry weight of its leaves), vast numbers of some species are collected to enable trace molecules to be concentrated into measurable amounts. In areas of high risk, this could be crucial in wiping out a population. A 'new' area of research is the coral reef, where unusual drugs might reside. In the early 1990s a US group investigated the Acorn Worm, *Cephalodiscus gilchristi*, taking 450 kg in order to extract only 1 mg of a potential anti-cancer chemical. A sea hare, *Dolabella auricularia*, yielded 10 mg of a drug from 1,600 kg, while 2,400 kg of sponge produced only 1 mg for subsequent testing. Stripping coral reefs in the name of science is surely as unethical and unacceptable as any wanton destruction that would bring the might of the green movement on the head of a multinational, polluting agency.

The search is not cheap – at least the subsequent checking and testing is expensive. Only about one in 10,000 chemicals prove valuable, and even then recovery of the initial investment may not be quick. On the other side of the coin, the US-based pharmaceutical firm controlling the products of the Rosy Periwinkle earns $100 million a year from sales, and the industry's plant-derived drugs are now valued at over $200 billion a year. Computer simulations and screening techniques steadily make identification of potentially useful organisms easier and cheaper, so the collecting schedule is likely to increase in pace.

Indigenous populations are, rightly, highly concerned with this stance. The Western world takes without return; it raises money in the form of aid for Third World development, then watches profit-motivated companies (however good the outcome of their research might be, they still answer to shareholders) rifle the world's gene library. If a company or individual took control of a source of another country's oil or produce without payment it would be classed as international theft. On this basis, developing countries are claiming that taking genes is no different. Added to that, why should they pay for the reimportation of drugs or genetically modified species which originated from their lands? Costa Rica, a tropical island with an immense biological inventory, has made a stand.

In 1991 an agreement between Merck (a New Jersey company) and INBio (a Costa Rican biodiversity research institute) pointed the way forward. Merck paid over $1 million, half of which went to conservation and the island's national parks while the remainder will help INBio train local scientists (and farmers, housewives, or anyone willing to collect specimens for pay) to analyse and develop their own resources. In return, INBio supplied Merck with a range of chemicals derived from the wild. It is a risk for Merck: it takes ten years or more for any realistic returns to be made from such investments, and in the meantime Costa Rica is free to sell its genetic heritage to other bidders. In addition, a royalty payment will be made on sales of drugs developed from its domains.

In 1992 the Earth Summit in Rio de Janeiro included a Biodiversity Convention. In this, amidst much discussion, it was agreed that countries should control their own genetic heritage. In the following year, Australia put forward strong views concerning the use of plants from the *Conospermum* genus, which only grows in Western Australia. Screening showed that an extract might prove useful in preventing HIV, which can lead to AIDS, from reproducing. The US developers and Australia, as genetic donors, agreed a deal which could be worth millions if the drug proves successful. However, while such covenants between stable countries are viable, other organizations may not be in a position to sign documents which will properly benefit the local human population. This is especially true in South America, where there is a vast store of genetic material yet to be tapped, but there is no assurance that government-signed deals will profit the real custodians of herbal knowledge: the indigenous people of the rainforests.

While these isolated cultures are perfectly able to coexist with nature, relying on the forest for medical cures, they have no words for 'patent' or 'royalty'; nor would they have a use for money, unless it were to 'purchase' the land which is rightfully theirs. Estimates state that approaching one hundred Amazonian tribes have already disappeared, and others are threatened. In 1987, as a result of contact by illegal miners, a quarter of the population of one tribe died from disease to which they had no defence.

In an attempt to place a value on trading with indigenous people, and thereby secure their lands and harvests, The Body Shop (a cosmetics retail chain) signed an agreement covering intellectual property rights with the Kayapo people in the Brazilian rainforest. In Peru, a Shipibo shaman teaches the old medical ways, travelling between communities, and a traditional medicine project has been established. It is hoped that the principles of teaching, of passing on knowledge, will now be enhanced and expanded. At the same time, the much advertised 'rainforest products', where local people are paid for their wares, are not always on solid ground. Brazilian nuts raise only a couple of cents in the dollar for their collectors, compared with sales on the New York market. In Britain, the import value of £4,000 of mahogany – perhaps only one tree – might mean a payment of under $10 (about £6) for the natives of Peru.

One approach is to boycott products known to exploit those people at the start of the supply chain, but care is needed. The money received by indigenous peoples has already introduced drastic changes in their way of life and, in any case, is the timber you purchase from a 'renewable resource'? How do you know? How were the fruits obtained, the fish caught? At every step there are difficulties in identification and in gathering reliable information; it is sometimes to the advantage of suppliers and traders to confuse such issues.

The extent of forest lore, and therefore the value of local people's knowledge, should not be underestimated. During searches for useful plants in the first years of the 1990s, two methods of collecting were compared: random samples for subsequent screening, and plants recommended by locals. The latter proved almost five times as effective in revealing chemicals of worth.

Such collections, conducted using native knowledge, are made by what are called ethnobotanists. From Tanzania comes a new species of tree used by people to cure toothache. The Mayapple of North America, also known as the Mandrake plant, was used by native Indians to treat warts and deafness and to kill parasitic worms. Today, it is used in cancer treatments with close to a 50 per cent recovery rate. Some 4,000 plants are in science's catalogue as potential contraceptives, with 260 of them now screened and confirmed as holding active 'no children medicine' ingredients. Tubocurarine, used in modern anaesthetics, relaxes muscles and comes from the bark of jungle vines, while pineapple stems yield an enzyme useful for treating thrombosis, the biggest single killer disease in Western culture. Even when drugs are known, better sources may be found. Searches revealed that British foxgloves, the first source of digitalis, produced smaller amounts of drug than *Digitalis lanata*, a related species from Austria; digitalis has yet to be synthesized and is still extracted solely from plants.

Ethnobotany is just one, albeit successful, approach to gaining 'new' compounds from the wild and searches closer to home still reveal the unexpected. The number of plants referred to in medical folklore can be extremely high. The Kayapo people of the Amazon utilize over a thousand plants, while plant extracts used in the traditional medicine called Kampo account for nearly half the prescriptions given by Japanese doctors to treat ailments such as hepatitis, high blood pressure, headaches, diabetes and constipation. Around the world, 85 per cent of traditional medicine is plant-based; that amounts to the sole medical treatment used by around 4 billion people. As an example of folk medicine which requires a re-evaluation, garlic has long been believed to aid digestion. This is not only correct, but garlic has also proved to be an excellent antibiotic. Allicin, its active ingredient, 'suffocates' harmful bacteria by depriving them of oxygen, and can be used to treat congestion, colds, infections, acne and breathing disorders; it also lowers blood cholesterol levels. An old folk remedy for treating the common cold is to hold two cloves of garlic in the mouth, one against each cheek, perhaps a solution that at the very least prevented the disease from spreading by discouraging other people from coming too close.

Nature's bounty is far-reaching: we use its products, and attempt to match its designs. As we learn more, it is being increasingly recognized that natural designs, tested by aeons of natural selection, can offer cheaper, more direct routes to industry and manufacturing. Suspension bridges, for example, are subject to the same tension and compression forces as the mammalian spine; supporting struts correspond to ligaments along backbones. Buttresses are formed by rainforest roots, models for efficient architecture. The first tunnel under the River Thames was built by Brunel after studying the shipworm's technique of lining the loose soil of its burrow with a hard material as it delved onwards. In mathematics and computing, the theory of natural selection is now used to model 'genetic algorithms' in solving problems.

One fact is certain: just as the search for genes reveals new medical options, retesting ancient remedies, from European and Chinese herbals and rainforest cultures alike, yields priceless drugs. Perhaps, though, the term 'priceless' is a naïve concept. In a similar fashion to Armstrong's footfall upon the moon, to extract a new drug from a poisonous plant may be a giant leap for mankind, but without care it could equally spell out a giant step backwards for man's genetic custodians. From whatever source the information is gained, it is truly time for Western cultures to pay the price for what they covet and value. And, if they covet and value biodiversity for whatever reason – ethical, moral or plain avarice – it is essential that Western cultures wake up to the devastating genetic losses occurring day by day in the rainforests and elsewhere. When species and their genes are lost, with them go the solutions to our questions.

CHAPTER NINE

THE SPICE OF LIFE

An investment in knowledge pays the best interest.

Benjamin Franklin, 1706–90

It is a misfortune frequently lamented that new truth, the most precious attainment of each generation, is also the most unwelcome. We do not hasten to sweep out our stock of laboriously collected ideas, even after the worthlessness of the assortment has been declared.

O.F. Cook, *Popular Science Monthly*, 1904

'Closing your eyes won't make it disappear.'
'Neither will talk.'

Discussion as a new road threatens the Amazon rainforest in the film *Medicine Man*, 1992

When Jonathan Swift wrote the satirical *Gulliver's Travels* in 1726, he took his hero, Lemuel Gulliver, through many lands. As well as Lilliput, Brobdingnag and Houyhnhnmland, Gulliver discovered Laputa, a land filled with sorcerers, scientists and immortals. By comparison, planet earth might be peopled by scientists and even by sorcerers, but neither its inhabitants nor the world we know are immortal.

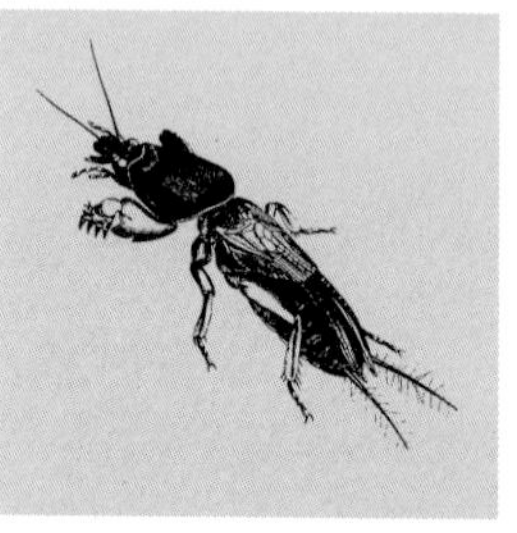

In the distant past, cells evolved in primeval oceans and began to feed. As they did so, releasing oxygen, the universe changed a little, then a little more. Oxygen promoted life, carbon dioxide trapped warmth: adapt and survive, change the world and adapt again, it was life that modelled our planet. Not just one species, but all of them, collectively. Make no mistake here: there is no implication that evolution or these alterations were part of a deliberate, thought-out plan – only that over 3,500 million years of life the net effect has been one of a changing environment, driven by the forces of life itself.

In 1972 James Lovelock, a radical, independent scientist, researching the possibility of life on other planets, was studying Mars. We now know that this cold planet was once much warmer (there are signs of rocks weathered by running water) and could have supported life. The problem with identifying whether it actually did so was the enforced, long-distance analysis: what feature of a planet might indicate, from afar, the presence of life? The answer which Lovelock and his companions produced lay with the composition of gases in the air: if it was 'unusual' (that is, unexpected for the planet's anticipated temperature or composition) then the atmosphere must be influenced by living things.

His thought processes in motion, Lovelock turned his attention back to the earth, seeing it as if observed from space. Earth's atmosphere proved an unusual one, a gaseous envelope which did not match predictions based on the planet's distance from the sun or composition of its rocks. At 150 million kilometres from the sun, earth's air should be approaching a temperature of 300°C, and be choked by carbon dioxide with a trace of nitrogen. Through past ages it changed its mix, peaking with oxygen at about 35 per cent at the time when insect life burgeoned. Today, earth possesses an 'abnormal' atmosphere of 79 per cent nitrogen and 21 per cent oxygen, laced with a trace of carbon dioxide and plenty of water vapour.

The oft-quoted fallacy that life arose on earth but not on Venus (hotter) or Mars (colder) because earth is 'just the right temperature' is therefore untrue. Lovelock's imagination developed a vision, wherein earth was a gigantic, self-regulating 'super-organism' which controlled its own atmosphere. Musing with the novelist James Golding over what to call the theory, Lovelock followed his friend's suggestion: the first words on his Gaia hypothesis were published in 1972.

Gaea was the personification of Mother Earth in Greek mythology (she was named Terra by the Romans), preceding Zeus and the gods of Olympus. She was both mother and wife of Uranus, together parenting the earliest living creatures. As a goddess, she can be cruel and caring: the earliest monsters were spawned by Gaea, while as Nature she determines with infinitesimal pain the fate of her domain. However, as Lovelock's idea became better known it seemed that the name he chose, Gaia, was unfortunate, for by inference it carried his scientific theory into the realm of philosophy and religion. If not a buzzword in the same way that 'biodiversity' has been adopted, Gaia has nevertheless prompted discussion and a different way of thought. It presented exactly what was needed: a name which captured the public's imagination

All populations fluctuate as factors such as disease, space, food, climate and predation have an effect. Ground squirrels might rise in numbers during a year when food is plentiful but, in a negative feedback effect, the following year decrease in numbers as their larger population provides a ready source of prey for animals higher up the food chain.

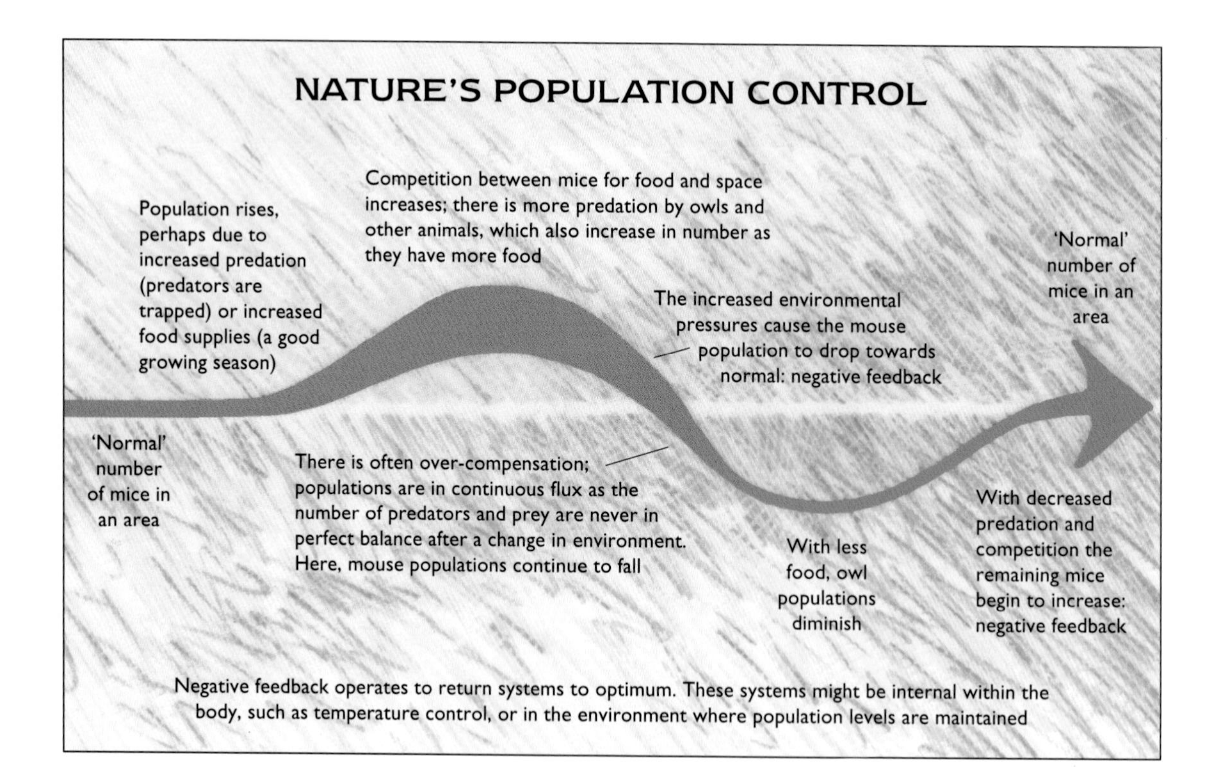

in a way that 'biological economics' or 'population dynamics' never could.

The concept of feedback is at the centre of Gaia. It is an old idea; a feedback system exists within all living organisms as well as much of today's technology. In essence, it provides a means for correcting an imbalance; of taking information and reacting to it. A bridge has too many cars upon it, so the barriers come down and the traffic flow is limited until the system returns to normal. This is the essential point: once the correction has been applied (limit the traffic), the result of the correction must be detected and the corrective mechanism ceases (the traffic can flow once again). This is what is termed a negative feedback: the corrective mechanism operates to return the system to a pre-set 'standard'. Detection, correction, detection, cease correction . . . When the opposite occurs, when a mechanism spirals out of control, it is termed positive feedback. The speed at which ice ages come and go, draining the land of water and then reflooding it as polar caps melt, can be explained by positive feedback: as ice forms it reflects heat back to space, which cools the planet further and forms more ice, which reflects more heat which . . . If some factor triggers melting, the oceans absorb heat which warms the planet which melts ice faster, and so on. To initiate such changes, a relatively minor temperature fall or rise is required.

Humans, as they are warm-blooded, produce and release heat; if too much heat is generated by the liver, the blood distributes it around the body and any excess is lost through the skin, returning the body's temperature to normal. However, if too much heat is being lost, the skin's blood vessels constrict and sweating stops, helping to retain body heat. In the same way, kidneys control the amount of water in blood, carbon dioxide levels are controlled by breathing and heart rates, and the liver adjusts blood sugar levels. If too much sugar is detected by the brain, it causes the pancreas to alter its production of insulin, which in turn causes the liver to release or store sugar. Such systems also arise outside the body: a rise in the population of mice is countered by more predation by owls as their food supply has increased, and there is added competition between the mice. There is 'environmental resistance' to the larger mouse population, resulting in a corrective, feedback mechanism. Feedback: a system of detection and adjustment.

In the Gaia hypothesis, earth's biosphere possesses such a regulating mechanism. To the theory's supporters (there are plenty of dissenters), Gaia has created, maintained and adjusts the conditions which permit it to survive. Just as with the theory of evolution, it is not a considered plan – only that Gaia has this effect due to naturally occurring, non-sentient controls. At its most basic, Gaia's control relates to earth's temperature.

Too cold or too hot, life will fail. Earth's temperature depends on a number of factors: the energy it receives from the sun, how much of this is absorbed, and how much is prevented from being reflected back out towards space. Of the first, the sun has steadily increased its output over the millennia so that it is now giving out about 20 per cent more energy than it was when the first cells formed in earth's oceans. To that changing input of energy, earth's life has responded. A green mantle of plants covers the earth, absorbing heat while removing carbon dioxide and releasing water vapour into the atmosphere. Both of these are greenhouse gases: they trap heat by allowing the sun's shortwave radiation to enter, but prevent its reflected longwave radiation from leaving. Carbon dioxide is removed from the air by organisms, which use the carbon for building cells. In particular, this involves microbes in soil: because of micro-organisms, there is about forty times the carbon level in soil as there is in air. Added to this reservoir of carbon is the amount locked in carbonate rocks: limestone, formed from countless numbers of organisms.

As life developed over the millennia it therefore removed carbon dioxide from the air, which, as this normally traps heat, implies a cooling earth. However, this effect is balanced by warming due to the presence of water vapour, pumped into the air by plants, which helps retain heat. Together, these factors form a balancing control which can adjust the effects of the sun's altered output: at the sun's lower energy level of 3,500 million years ago, today's oceans would freeze.

More than that, the temperature has been ideal for the past development of cold-blooded organisms: warm blood is a recent evolutionary phenomenon. Add to that a repair mechanism which replaces life lost during the great extinctions: the new-for-old species mould themselves to an earth changed by sudden catastrophe, and then mould their planet still further. When these ingredients are taken together, the argument for Gaia's existence becomes strong.

Form a Gaian hypothesis: slowly increase the heat reaching earth. Life is then promoted and plant species grow faster, therefore using more carbon dioxide and reducing earth's heat retention system. However, at first sight there also appears to be a positive feedback system in operation: more heat, more evaporation, more water vapour, more heat trapped, more evaporation . . . However, as well as releasing water vapour and oxygen, algae in the oceans also produce a sulphur-bearing gas (dimethylsulphide; algae pump more sulphur into the atmosphere than the combined output of all mankind's power stations). In the air, sunlight converts this to an acid. Water vapour will coalesce on particles in the air – in other words, the acid provides a nucleus which water vapour condenses upon, forming small drops of rain. Oceanic plankton is therefore involved in the formation of clouds and, from space, this is the equivalent of a huge, white reflector which bounces the sun's radiation back into space before it reaches the planet's surface. The hypothesized, added heat from the sun has, in the end, led to a situation where the temperature of earth is kept at a constant: life has maintained its own.

So, if Gaia really does exist and Mother Earth can right all these wrongs, why all the furore about the greenhouse effect, CFC gas release, damage to the ozone layer, pollution, decimation of the rainforests . . . ? The answers are perhaps obvious: we *don't* know for sure if a Gaian system is operating; we only suspect it. How can a scientific experiment which requires two identical worlds, one as a control and the other to initiate experimental changes, ever be conducted? And, of course, as far as blind Gaia is concerned, the planetary whole is what is important, not one species upon it. Gaia does not possess choice: an attack upon the welfare of earth triggers corrective mechanisms, no matter what stands in the way. Nearly a century and a half ago, Charles Darwin wrote:

> *Man selects only for his own good: Nature only for the being which she tends. The slightest differences may turn the nicely balanced scale in the struggle for life, and so be preserved. How fleeting are the wishes and efforts of man! how short his time!*

In the face of Gaia, man is not an essential part. Far from it. Step on an ant, or exterminate a species, and the world still turns. When they go to sea to save those in distress, rescue services use a self-righting lifeboat. If it turns over in a violent storm, like Gaia it will roll back, buoyant and seaworthy. However, there is nothing to guarantee that every member of the lifeboat's crew will still be aboard. And, occasionally, the lifeboat is wrecked.

The much-heralded problem of the greenhouse effect is stimulated by two sources: wanton destruction of forests which remove carbon dioxide, and excessive burning of fossil fuels. Since the mid-eighteenth century (when industrial processes evolved) there has been a 30 per cent increase of this greenhouse gas, and the past fifty years alone have seen an average of over 0.5°C increase in global temperatures. Only half a degree does not seem much; certainly not enough for a person to detect from day to day. Doom and gloom predictions are common, so why the emphasis on such a slight change in climate?

Predictions can only be based upon what we believe to be the truth, then extrapolating data into the future. This shows that, if planetary warming continues, even minute increases could prove catastrophic, and 0.5°C represents the greatest warming since the last ice age. Weather patterns will become unpredictable, with more storms in some areas and less rainfall elsewhere: an increase in the world's deserts is indicated. In turn, agriculture, dependent on a stable climate for its managed crops, would be seriously threatened. Monocultures are poorly placed to withstand severe storms, and their genetic structure does not permit a biological response. As warmer weather predominates, earth's ice caps can be expected to recede (the effects of warming are concentrated towards the poles), their water adding to the oceans. A large quantity of methane is trapped in the ice of the northern tundra and, as this is a greenhouse gas with some 25 times carbon dioxide's ability to retain heat, the problem will spiral. The prediction model shows that even a small temperature rise will lead to flooding; huge areas of the globe lie at low altitudes, and a few centimetres of water would extensively contaminate agricultural land with salt and destroy freshwater habitats.

Biodiversity would pay the biggest price in this scenario. Plants, and the animals which depend on them for food and shelter, are frequently restricted to a specific range of conditions: their biological island parameters. We already set aside areas as game and nature reserves, some small and others large, but all neatly managed for the future. Biodiversity is preserved. However, change the conditions and a species' natural response is to migrate or disperse towards another suitable area. As temperatures increase, the zone for ideal growth for many species will move towards the poles – perhaps by several hundred kilometres over the next century. With life dependent on fenced and city-locked nature

Opposite: As we burn fossil fuels we add pollutants to our planet, increasing the levels of sulphur and carbon dioxide in the air. While the greenhouse effect is partially but precariously balanced by the losses to the ozone layer, we may not see the true effects of our actions until the see-saw effect is lost and Gaia can no longer offer her compensating system.

reserves, those species which cannot adapt or move with the climate are doomed. As reserves fragment, their areas diminished, species will continue to be lost. With a climate change of even a couple of degrees, today's biodiversity will certainly not be preserved.

When the search was made in the Andean mountains for a new species of potato, locating instead what became a genetically valuable tomato species, the area was chosen with care: it was not a random attempt. The tomato's discovery came as a surprise because, as specimens in Mexico show the greatest variation in shape, size and colour, it is believed that the plant was first domesticated in that region. A search of Ethiopia for pea plants, based on the fact that there are more known species in that area than elsewhere, revealed a species which conferred resistance to fungal infestations of pea roots. After viral attacks on US crops prompted a deliberate search for new species of barley in 1963, Ethiopia also revealed a plant which conferred resistance due to the presence of a single gene.

Other searches have been less successful. *Theobroma cacao,* which grows in areas now under examination for oil reserves, is the source of cocoa. Crops of the tree, grown commercially for their beans, are susceptible to viral and fungal diseases and, although some 2,500 strains of the species are known, no useful wild gene which will help with the problem has yet been found – but the genes may nevertheless exist.

In the 1920s and '30s a Russian geneticist, Nikolai Vavilov, made a series of plant collections in Central and South America, Asia and China. He logged the number of species related to crop plants in any one area, linking similar numbers with each other on a map to produce a series of concentric lines he referred to as isoflors; the pattern appeared much like height contours on a geographical map, or isobars on a weather chart. Vavilov found centres where crop species were the most varied, with outflowing tongues. He hypothesized that the centres of diversity corresponded to the ancestral home of the plant, while the tongues indicated regions where subsequent colonization had taken place. Outlying peaks of species showed where plants had colonized the land but had then been cut off from their parental population, allowing them to speciate further.

These species-rich regions are now called Vavilov Centres. They broadly correspond to such areas as the Horn of Africa, northern Arabia, parts of South-East Asia and Indonesia, central Chile, the Mediterranean, Mexico, and a zone stretching across Ecuador and Peru. In effect, these tracts of land proved richest in the species which man has cultivated, and logically are the first ones to turn to when seeking new genes for crop species. In effect, these were the regions of the world where civilization – indicated by the first domestication of plants – began.

A similar concept indicates areas of highest biodiversity, based on the sheer numbers of identified plant species rather than restricting the map to ancestral food plants. More than that, the species mapped are usually indicated only where they are unique: if they are common across the globe, the species are left out of the equation. Where the concentric lines draw closer, biodiversity is rich and also globally scarce. The atlas shows areas of vitality that we call hot spots (the opposite, species-poor areas, are termed cold spots).

Some hot spots coincide with Vavilov Centres, but there are many others. Fynbos (a form of heathland), found in the Cape Province of South Africa and in south-western Australia, is coated with rare flowers, while forest remnants are resplendent in central Africa and the isolated island of Madagascar is rich in endemic species. All are threatened to a lesser or greater degree: fynbos from agriculture, forests from logging. Areas of Indonesia, Malaysia and the Philippines have been and are being decimated by clear-cutting, whole islands stripped of cover in a short-sighted, localized scrabble for cash.

Hot spots can also be plotted for animals. For example, an analysis of bird species shows that 20 per cent of all identified species (and 80 per cent of endangered species) are restricted to only 2 per cent of the planet's land area. If hot spots for groups of species are plotted, such as for mammals and birds, overlapping areas indicate doubly-rich regions, though some groups will never coincide as their habitat requirements differ. Such studies enable conservation groups to target areas which will yield the most benefit; in the case of birds, Indonesia is easily top of the list. If global warming continues, these crucial hot spot areas will be further stressed: genetic alarms should, by now, be roaring in everybody's ears.

Many people believe in the inherent right of life to exist, but altruism is unlikely to convince a government earning the major part of its country's income from logging rights, any more than it might a giant conglomerate which has a financial stake in exploiting an area. 'Save the world' is all very well, but a race between decimation and preservation is involved. Realistically, what is required is a reason for the conservation of life which will present an inherent, measurable value to the accountants who cannot see beyond a profit sheet, yet hold the power. Habitats must pay their way and hence force their own protection. Rainforests must turn commercial.

Maintaining our existing crops and developing new ones requires a rich genetic stock; because Neolithic farmers relied upon more species than we do today is no reason to continue in the same vein. The 'supermarket on a stalk' plant, *Psophocarpus tetragonolobus*, has often been introduced as a contender for a 'new' food. It grows quickly in poor soil, and all parts of the plant can be eaten. The high-protein beans can be made into flour or a late night, coffee-like, caffeine-free drink, the roots can be eaten like potatoes and the young seeds like peas, while the leaves can be boiled like cabbage. Waiting to be brought to our attention there are new spices and, in particular, fruits. For protein the 'supermarket plant', and other species, match or excel our current crops – including the ubiquitous soya bean. Other plants manufacture sweet compounds which place sugar beet and cane in the shade. There is no need to rely upon saccharine: the 3 m high *Thaumatococcus danielli* produces a seed coating which exceeds sugar's sweetness by 4,000 times. The rainforest already gives us ice creams based on fruits, as well as varnish from trees, cosmetics from flowers and body oils from seeds. A brewery in São Luis burns the shells of babassu palm coconuts in its boilers, while Brazil's automobiles run on alcohol made from sugar cane.

Such forests products are true renewable resources: nuts grow again on the same tree, fruits can be plucked by the land's custodians next season. Some Brazilian trees are pollinated only by female orchid bees, which means that without a supporting ecosystem plantations will fail; the commercial harvest of these nuts, worth $50 million per year, *must* come from the ancestral home. In pharmaceutical values, it has been estimated that one in every ten plant species may contain compounds which are active against cancer; what price the rainforests if the cure lies within? Even minuscule details are commercially important: if your butter contains an orange colouring agent, E160b, this is made from oriana seeds whose ancestral stock came from the Amazon. With care, sell the products and save the forests.

The list of such benefits is endless, the more so as we are incredibly ignorant about our world; we hardly know what we don't know. As yet, of the 275,000 plants which have been identified, only about 10 per cent have been investigated for their value to man (and animals virtually not at all). The upper estimates of all living species used to be placed at around five million species. Then ten million. Each plant, on average, supports up to forty species of animals, and these estimates are based on what has been identified so far. Naturally, larger organisms are easier to locate and identify, while smaller species are more numerous but harder to locate. In all, we have catalogued a minuscule proportion of life, only 1 per cent of which is larger than a bumble bee.

In 1982 biologist Terry Erwin stepped into a South American rainforest to conduct a survey of its insect life. Using a fogging technique, essentially blasting a volatile pesticide into the vast, inaccessible treetop canopy, he and his colleagues collected what tumbled out. From a single tree Erwin obtained 163 species of beetle. By extrapolation, multiplying his findings by the number of known species of trees and percentages of beetles to other types of insects and spiders, Erwin arrived at the staggering total of 20 million species of 'bugs' in the rainforest canopy alone. Mathematicians refer to a small change in a formula which, originating early on, can lead to a later, disproportionate change. It is called the butterfly effect: philosophically, the flight of a butterfly might change the course of an air current, and tip the balance towards a breaking storm. It is relevant here: change the number of Erwin's beetle species and the final estimate could be drastically higher or lower. However, considering that there are also the species of the rainforest floor to add to the total, plus those of the oceans and other habitats of Gaia, thirty, forty or even one hundred million species might be currently dwelling alongside man. Pick up a couple of Brazil nuts: an equal weight of soil might contain a million microbes, and who knows if one is new to science.

One answer to the diminishing global gene pool has been to collect seeds as rapidly as possible, placing them in gene banks against a future need for withdrawal. This 'quick and dirty' approach is notable for its almost panic-like requirements, and some species are proving recalcitrant: they do not survive unless grown quickly from seed. The cocoa family of trees forms one example: if the seed is not quickly planted, it dies. To maintain the stock the trees must be preserved, rather than the cocoa bean, meaning that a cocoa plantation on Trinidad must grow specimens of all 2,500 strains. If, instead, the rainforest home of the principal species *Theobroma cacao* was preserved, how much better for the chocolate industry: chocoholics take note.

Gene banks apply not only to seeds: frozen sperm, ova and embryos are held as genetic safeguards, but for reasons of cost and practicability this is only useful (if successful at all) for helping preserve the earth's more spectacular life-forms: the leopards and tigers, the large 'appealing' mammals. While the genetic material contained in the world's 'bugs' is more vital to biotechnology, its total salvation is an utterly impossible task for a gene bank, but there is at least a slim hope for the zoos of the future. What is devastatingly sad is the potential loss from the wild of the Indian Tiger and Javan Rhino (each numbers under a hundred, and poaching continues), and of the dolphins and whales. Plunge their cells in liquid nitrogen, wait and wonder if there is a tomorrow. Whatever means are used to preserve the genes of plants and animals – gene banks, zoos, botanical gardens – they can only encompass a minute fraction of the whole. Even if the approach is valid, a gene bank will never buttress the covenant which our species owes Gaia. It is an exceedingly poor Noah's Ark.

It is relatively easy to maintain support for the cute 'n' cuddly species, the spectacular giants which draw eco-tourists and their cash. Now, it is recognized that campaigning for the endangered few (no matter that there are thousands of species on the list, they are still a minority to Gaia) is less valuable than conserving whole habitats. Without a sheltering home, a species will never prosper, but the 'wild' portion of our wilderness is disappearing fast. Yet, the warm, furry and colourful provide a rallying point, even though mammals only total a quarter of one per cent of Gaia's life. It is not too far a step from saving an owl or beautiful butterfly to conserving the forest or heath habitat it requires. If a spectacular species is the centrepiece, so be it. Let it be the speaker to the unconverted. Planet earth gains, as well as its species, for habitat diversity is crucial.

In this way the endangered Spotted Owl (*Strix occidentalis*) proved the turning point for the fight to preserve the forest remnants of the USA's Pacific North-west. In Britain the Natterjack Toad's diminishing numbers conferred protection to heathlands and wetlands. The Norfolk Bittern, a bird once so common that it was hunted for food – it was

This four-month-old baby Sumatran Orang-utan (*Pongo pygmaeus*) was born at Bristol Zoo, England, helping to demonstrate the part that zoos can play in conserving the genes of the wild. The question remains, however: do we wish our wildlife to rely on cages, or the realistic protection of native habitats?

apparently excellent when roasted – decreased to only 15 pairs before its reed beds were conserved and replanted. Locally, as hedgerows are destroyed and pesticides and road-building increases, we destroy faster than we can save: the Norfolk Bittern survived and its population is increasing, but Britain contains more than 1,200 other species under threat.

Rainforests and reefs are always high on the biodiversity agenda, as they possess so many species for their total area: less than 8 per cent of earth's surface. Playing the integration game, however, forests and reefs are interlocked with wetlands and desert, mountains and tundra, lakes and seas, and each is an important part of the whole. Mangroves grow with over twenty times the ocean's productive potential, with reefs, estuaries and shallow algal beds almost as rich. Over a third of all species of fish, more than 6,000, are found on reefs and, though these only occupy 0.16 per cent of the earth's ocean area, they sustainably yield 12 per cent of the world's total seafood harvest.

Forest habitats carry half the world's biomass, but also cannot survive in isolation. The rainforests are the Gaia's powerhouses: their hot spots are truly unique and critical to Gaia's functions. Yet, these are precisely the areas where timber mining (what better description for the wholesale clear-cutting of natural forest?) proceeds apace. We are used to hearing the rate of forest loss: huge, meaningless numbers. Global forest clearance and decimation now

The Indri, related to the lemur, lives in Madagascar and is the size of a four-year-old child. Its call, reminiscent of whale song, has produced the superstition amongst the locals that it is a human in another form. This has literally saved its skin from hunters, though as forests are felled its habitat will become reduced.

amounts to over 200,000 square kilometres each year. This is the equivalent of the area of Great Britain (which has already destroyed virtually all its native forest) every year, or three times the extent of Washington DC per day.

Over the last forty years the earth's rainforest cover has been halved, and the rate of loss steadily increases: at the current pace, the last Brazilian tree will fall in under thirty years. Ghana has already lost 80 per cent of its rainforest, and Madagascar a similar amount with the remainder consistently burned and degraded. Brazilian coastal rainforest is now down to 2 per cent of its original size. Twenty years of logging has removed half of Ecuador's forests. The story is the same the world over:

Vietnam, defoliated with Agent Orange and cut by farmers, to 50 per cent forest cover; Thailand reduced to 25 per cent; Nigeria to 10 per cent. Madagascar and Nigeria are due to mourn their last trees within ten years, unless the losses are stemmed. These estimates ignore the effect of replanting, of course, but there are insidious dangers linked to statistics which merely record the area of forest. When ancient trees are felled, with them go their dependent species. Replacement forests often consist of alien plantations, as is happening on Madagascar. Native faunas will not, of course, find their food supply in the toxic leaves of the

Rainforest losses are increasing. New Guinea is unique in having retained 70 per cent of its primary forest, though moves to speed up the rate of logging have begun.

usual eucalyptus forests, and biodiversity suffers a further knock-on effect.

It is not so much that rainforests are vital for producing oxygen (the sea's plankton is the single greatest producer), but their trapped carbon is released in surges when the trees are burned – fuelwood is not only the ultimate end for a felled giant, but the actual intended, immediate use for over half the felled timber. Living forests are sinks for pollution, they are rainfall recyclers, cooling agents of the air. What will happen as they disappear? Theories of global warming once indicated that, as temperatures rose, rainforests would emit more water vapour and there would be more clouds and more rain. In 1993 the Meteorological Office in London ran a computer simulation which showed that, in fact, there would be less rainfall over huge areas as the forests initially grew faster: they would *absorb* more water than they released. In turn, the trees impoverished of rain, forests would give way to grassland before the balance could be restored. It is a dynamic state: change one factor and the dominoes fall, but it is hard to tell in which direction.

The difficulty lies in the trigger which man is squeezing: global warming is not affecting the forests in the way we anticipate, because we are confusing the issue by stripping their resources. Gone, the trees' water-and-light sponge effects lost, soil can wash to the sea and clog rivers. Removed, forest leaves no longer absorb the sun – and here is another anomaly. With no forest to cover the land, the earth's albedo – its 'shininess' – increases. It reflects more light back towards space, a different cooling effect. Gaia playing dominoes again.

We have great difficulty in accurately predicting what the future holds; Gaia is incredibly complex. Like a Russian doll we strip away onion layers of fact, but never reveal the kernel of whole truth. We cannot accurately foretell the weather next week, let alone next month. To give an example: predictions show that the half-degree temperature rise attained over the past century is not enough of a response to the pollution which man has already introduced. Aspects of the existing rainforests cool the planet, but elements of their loss do the same.

In an attempt to switch from burning fuel for energy to one of harnessing energy from water power, hydroelectric dams are being built. When the Amazonian Tucuruí Dam was constructed it flooded vast areas of vegetation. Research conducted in 1995 shows that, as the plant material rots, it releases methane in significant quantities; it contributes 26 times the effect on global warming than a coal-fired station producing the same power. Wherever a solution to a problem is tried, the result may conflict with the aims. However, in Gaia's terms feedback may already be operating – to a point. It seems likely that the damage we inflict may be partly masked by other factors, or that the damage conceals harm created elsewhere.

The 'ozone hole' is a real misnomer. Earth's upper atmosphere naturally contains ozone, which is broken down by man-made CFC (chloro-fluorocarbon) gases. The effects are especially concentrated towards the South Pole and, while an actual 'hole' can occur in the ozone layer, for the most part the layer only thins rather than disappears. Ozone traps heat, and so helps warm the earth; thus, in effect, ozone loss operates to reduce the effects of global warming.

This is not necessarily good news, as ozone depletion has other, far-reaching effects. Ozone also acts as a shield against high-energy, ultra-violet radiation from the sun (UV-B), the same source of energy which produces a suntan in human skin. Ozone losses are currently about 2 per cent each year, and the added radiation reaching earth is powerful enough to damage fragile plankton cells in the sea and increase the incidence of skin cancer. Human eye problems are expected to increase over the next decade, and sheep with cataracts have already been identified in Chile, an area which often lies beneath a low-ozone area.

Biologists use 'indicator species' to determine the state of habitats: sludge worms signify polluted waters, while lichens indicate clean air. Amphibians also work as global indicators of harm, responding quickly to adverse change: Australian and US species are declining, while orange-coloured frogs in Britain seem on the increase. Explanations indicate a link with radiation increases and global warming: spawn in shallow water is easily affected and killed by ultra-violet light, while the loss of a tadpole's normal dark pigment (if it survives that long), leaving it orange, makes it better suited to development in warm water.

Although CFC gas production has now decreased, other pollutants have the same effects, including some fungicides sprayed on crops. Additionally, as with global warming, there can be unexpected results. The Antarctic is rich in bacteria which feed on the algae that emit sulphur-bearing compounds, which in turn aid the formation of clouds. While global warming increases algal growth and hence more clouds to reflect heat back to space (a negative feedback), the reduction in ozone permits more ultra-violet radiation to reach the earth. This reduces the activity of bacteria by as much as 40 per cent. Thus, an ozone reduction will increase the algal plankton and help reduce global warming by opposing the effects of greenhouse gases; furthermore, simply having a thinner ozone layer will permit more heat to be lost. The twin problems of ozone depletion and global warming are both likely to continue in opposition; with such conflicts in cause and effect, it is difficult to see past the mask and chart the outcome. Interactions which span the globe are still beyond our best computations; less money is

Lichens are used as an indicator of clean air, as in this instance of a rock-dwelling species on Skomer Island in west Wales.

spent on ecological study than on the arsenal of some small countries. We know more about genetics than evolution: we create and destroy, but do not know the effects of what we release.

What is clear is that we cannot continue to think of our energy needs as coming from dwindling stocks of whales or timber, as if from a mine. When the minerals are removed from soil, there is no regeneration. When a rainforest is gone, just as when the last whale sings its song, there is no replacement. Some forester scientists will challenge that statement: rainforests, like any other habitat, are in flux and will regrow, increasing their diversity as new niches are opened by localized damage. True, but only when enough of the forest remains to regenerate from. That is not something we are permitting.

Homo sapiens has shown a propensity for destruction which is unparalleled in the history of Gaia. Where Gaia can self-repair, man can negate the opportunity like a blight. In global terms, we are a plague, tearing down the healing mechanism; imbalances are now set into Gaia's system. The fear is that we will bring those imbalances to a point where there is a sudden crash: Gaia's mechanism will switch from one of protective, negative feedback to one of reinforced, positive feedback. The system fails, and each element of Pandora's box adds fuel to a spiralling wave of cause and effect. Eventually, there will be a straw for the camel's back – a butterfly effect to tip the balance. CFC gases mask greenhouse gas effects, but do not survive forever and there may be a sudden imbalance on the horizon. For Gaia, there may be a critical point awaiting, beyond which there is an inevitable storm. The question is not whether Gaia herself will survive, but in what form she will do so.

We are forcing species to extinction on an hourly basis, at perhaps five hundred times the rate which it occurred at in our immediate past. This needs to be placed in context: we are losing species at an even faster rate than during the mass extinctions. Not that we can be sure that a species is extinct without watching the last specimen die in captivity. Might there be more in a remote jungle, waiting to be rediscovered? The Coelacanth, an ancient bony fish, was thought to have become extinct at the end of the Cretaceous, only to be caught in a fisherman's net in 1932. Another specimen was captured off Madagascar in 1952; somewhere in the oceans, this relict lives on. We so poorly know the rainforests that even large mammals are escaping our attention; as recently as 1992 a new species of monkey was found, a 'flagship to our ignorance' noted Russell Mittermeier of Conservation International.

We are not certain what triggered the mass extinctions, only that we suspect that there was a sudden catastrophe from which Gaia shuddered, staggered, and repaired herself. This time round, though, it seems we cannot look to the skies and watch for comets. The heavens do not need to provide the killing blow. Mankind, in a subtle, suicidal pact, is bringing on the sixth great wave of extinction. *Homo sapiens* is a force alongside the cyclic ice ages and continental drift – only with the talent for speed added to that of immense change. When man vacates the earth, it will be due to forces under his control.

There is no easy or obvious panacea. Our populations grow, demanding more resources. When a physicist increases the molecules of gas in a sealed globe, activity burgeons. Collisions multiply and temperature rises. Humankind is no different: with greater numbers, our freedom of choice diminishes at the same rate as our problems increase. A limit to our population must come. When yeast respires in a sealed flask it reduces sugar to alcohol and carbon dioxide. More sugar, more alcohol: brewing is an ancient discipline, but with limits. There comes a point where sugar just sweetens the wine; above 17 per cent alcohol, the yeast dies in the product of its excretion, its environmental pollution coveted as a drink by

Opposite: A symptom of man's attitude to nature and education. These dyed, dried starfish lie forlorn in a basket to be sold to children for a few cents each. The public aquarium which chose to stock its shop with carved shells and dyed skeletal remnants of coral has won awards for its strong conservation messages.

An industrial pipe spews pollution into a river not far from the British coast. We cannot afford to continue polluting our planet in this way; Gaia's abilities are not limitless.

another species. There are critical points for humans and Gaia alike: we cannot pollute continuously without eventually finding the alcoholic threshold; Gaia will change and survive but, having forced her hand, we are offering ourselves an existence with reduced choices.

The damage is being done, daily. There are abundant reasons for saving Gaia's biodiversity; this book is full of them: commercial, philosophical, ethical. The threats are easy to identify. The solutions are harder to come by. How hypocritical is it for our planetary leaders to bemoan our losses with a whisper, satellite-track earth's damage and construct elaborate 'endangered' reports, while smilingly signing licences and import orders which lead towards devastation? That we demand protection of 'foreign' countries' species for our uses, while polluting the air at their expense? That we campaign against rainforest destruction, but purchase the timber for our stairs, bookshelves, toilet seats? We market the idea that biodiversity is commercially valuable, that desirable life is worth protecting: catalogue the world to find the nuggets of gold. In the future, will our children thank us for a different problem: how to stop the destruction of what does not appear on the list, what has been deemed useless? Where will a profit-motivated ideal take us – to the realm of eco-tourists, paying to witness the vestiges of wilderness? As to 'civilizing' the people of the rainforests, we crush them into smaller homelands as the trees are stripped yet, two-faced, we demand a price: their genetic heritage, their oral traditions and knowledge. Rainforests and oceans, our green planet has become a political football.

Consider the worst war: the first strike is over, rockets and shells have pounded their targets from afar. Armies march, sickened by what they see, blindly following orders. Civilians feel helpless. The occupation of a devastated land is complete. Two things have been destroyed: the physical structure, and that indefinable something which a conquered race cannot recover. With the effort and will of the leaders, architecture can be rebuilt, the cannon craters filled, but it will never become as before.

We have one planetary ecosystem, the only biosphere we can survive in. We are doing those same two things to our home: destroying on two fronts. In the end, with the

Opposite: Victoria Falls, on the border between Zambia and Zimbabwe, is a Mecca for eco-tourists eager to witness the forces of nature at work. Without protection, such magical places will not survive.

will and the effort, we could theoretically rebuild a forest, redrain our flooded marshes – but with no ancient, genetic heritage. If we wish, pollution and population *are* restrainable; with study, we *can* learn to control. But Gaia's species, her biodiversity, once gone are lost forever. Evolution will replace them with the same or greater variety, but 10 million years is a long time to wait. There is no turning back. Sceptics say, 'The facts are wrong, the premises are unsound. There is no risk.' Why should those who believe the danger, or who are even unsure and undecided, be forced to gamble their children's future? This is too imperative a course of action not to take the safest course. Only a sick and troubled person plays Russian roulette: sooner or later, the chamber is full. With Gaia, we may never hear the gunshot; the target's evidence may be revealed too late.

In 1871 Alexander Agassiz wrote of the problems bestowed on biology by its division into separate departments. Anatomy and physiology were the children of medicine, while geology gave birth to palaeontology. He argued the need to unite them all, psychology, embryology and the rest, with a common link under one roof of study. If not, the laws of nature 'must remain unintelligible to him who is busy with naming and classifying materials . . . merely accumulating facts to be stored in museums, forming as it were a library of nature. To him its books will be inaccessible and its laws as inexplicable as are the laws of the motions of the planets to one who has no knowledge of the existence of gravity.'

Over a hundred years later a similar ethos might prove prophetic; we need to understand more of the vastness of nature than the narrow categories of commerce, economics or agriculture. We must tackle these and wider issues. We have waited, with warnings, for a hundred years. There is no longer the time left to delay. There is a problem. In the words of Helen Keller, 'Science may have found a cure for most evils; but it has found no remedy for the worst of them all – the apathy of human beings.'

What will be the outcome of science's study of biodiversity, a word which encompasses not just an array of species, but also communities and habitats? Today, there are fewer taxonomists cataloguing our world since Linnaeus's time – there is simply not enough emphasis on understanding the complexity of our world. To a planet of mankind, the promise of our rainforests comes to naught unless there is widespread understanding of its plight. To hold information in a catalogue is meaningless, like an unused, dusty library; what we learn must be put to use. The war is in session, and Gaia is in retreat. Those who understand this must choose sides, and be prepared to do more than watch from scientific sidelines.

Of where and how life began, we have an inkling; where life has ceased, we can trace and surmise: we think we have captured answers. As to where life is heading, of that portion of Gaia we have now taken control. Whether we have the humility, the will or ability to act wisely on behalf of the Goddess, that is another question. The future is plastic; we can mould it to our will, but only as branching choices from the point we occupy today. At our present rate, a quarter of all species will be gone in the next fifty years. We might be able to stem the loss; we are certainly capable of greater destruction, though we plead with eloquence.

Where will we go, in our decimated future? John Donne's poem asserts that 'No man is an island entire of itself.' Famously, it continues: 'any man's death diminishes me, because I am involved in Mankind; And therefore never send to know for whom the bell tolls; it tolls for thee.' We are inextricably bound with Gaia; be sad not for the rainforests and reefs, but for what our planet jewel and our own existence has become.

Our voyage of discovery, our association with a biodiverse universe, continues. What is our moral imperative? We control life. We create life. We, the evolutionary newcomers, seek knowledge and power.

What *are* we going to do with it? Think wisely before you answer.

EPILOGUE

TODAY, TOMORROW

I AM SITTING, WARMING at a wood fire, surrounded by wet, lush grass and a deep, star-filled night. There are others here, too, quiet except for the crackle of twisted mahogany in flames. The wood was collected miles distant; there is nothing dry here in the swamp surroundings. The ground is wet, verdant grass already soaked with night mist which will stretch through morning. Trees, vine laden, droop from higher ground, ranks of sentinels in the glimmering light.

Saiwa Swamp lies in western Kenya. I hadn't expected to be cold and wet, here in equatorial Africa. Well, cold at night perhaps, but not wet in weather more reminiscent of an English autumn. I've just come from the elephant caves of Mount Elgon, and before that a remnant of rainforest at Kakamega. Kakamega is like an insulated island; atoll-like, in the midst of open veldt and farmed soil surrounded by a belt of tea plantations and bone-cracking, organ-churning rutted roads, it stands proud and unexpected

Kenya's Rift Valley lakes are rich in salts and warmed by thermal vents, providing ideal feeding ground for flocks of flamingos. It is such wild places as this that earth's wildlife depends upon for its last havens.

with a welcoming wooden forest station where visitors can sleep. Once, a great forest lay across the continent, stretching through Zaïre into the interior. This remaining stand of tall, indigenous trees is all that is left, filled with the screech of monkeys and a flood of blue turacos.

Saiwa is different. Few visit here, leaving the treetop vantage points empty for those who wish to behold the sitatunga. I spent a whole day, perched in dead branches, before two of these aquatic antelopes waded and swam through the swamp before my camera, delicately grazing on waterweed beside otters and cranes. In Saiwa, flowers bloom with careful patches of gold. Spiders spin intricate webs, always glistening in the ever-present mist which hangs on the threads. Kingfishers flash across the water; bushbuck clatter through jungle. Like Kakamega, the swamp is special, the more so for its solitude.

The burning wood dies towards thin, grey ash, occasionally drifting glowing sparks up through the sodden trees. A chance to think, prodding the embers with a stick.

Why was I here? There's the obvious: a chance to view nature far from home is treasured. Scattered memories match the isolated ecological islands, still rich after time. It's an escape from computer screens and petrol fumes, from anonymous people and noise. Personal reasons that are hard to define, but interlinked with a need to see what earth's biodiverse store still offers. To be there, to feel it, to experience first hand. Most things we do, when it comes down to it, are as hard to explain.

'*Kwa nini?*'

Earlier, shortly after the sudden nightfall which Kenya brings, an African man came out of the night and squatted by the fire. '*Jambo*,' the universal Swahili greeting. His hands were thrust toward the flames to soak up the heat, their lighter coloured palms shining in the dark. He pointed to the camera; he'd seen me looking for the sitatunga, carrying a bag of lenses. '*Kwa nini? Why?*' He wanted to know why I made the effort, why drag this weight of metal with me, why look for this animal? His English faltered, but was good. What was the point? He didn't understand.

How can you explain such needs? I was facing someone who had lived with the land all his life. He knew the animals and plants surrounding him in a way that I never will, no matter that I've travelled afar and seen so many sights. What could I know of him? I was in his land, where children point and cry '*Mzungu*' – white face – and he spoke to me in my language whereas I was ashamed to know so little of his. His way of life depended on such fundamental factors as living in a successful community: would the rains support the crops, would the cattle produce a calf, was there food? He could not see the need for this alien thing, a photograph, which all tourists took. Yet, to me, to photograph within a natural world is as vital as his knowledge about the land and what it provides.

He squatted, and then sat on a fallen trunk, head tilted to one side. I thought of answers.

'*It reminds me of where I've been*,' I stumbled. '*The pictures, like a diary.*'

He looked more confused. Why should I need to remember such things? I will live as well tomorrow without them, as with. I tried again.

'*People, friends, when I get home, will be able to see where I have been. They can share what I have seen.*'

That brought a little more success, but not much. A different tack:

'*We do not have trees like these. We have no monkeys or elephants in our land. Our home is different; different flowers, different plants, different animals. No hippos, no buffalo, no hornbills. Friends can see what they look like, with a photograph.*'

The elephants caught his attention. Everyone came to see the elephant. Had we come to see the elephant? '*Yes*,' we replied, '*ndiyo: it is so.*' He was a little happier, as he had a link: we came for the elephant, we brought our money; as in Namibia and Zimbabwe there was a comprehension that the animals and land drew the tourist, even if he could not understand why. I asked him what he thought of the elephants, were they not magnificent?

He shrugged, somehow a curious gesture, perhaps just due to the cold: the fire warmed from the front, while the night froze from behind. He had never seen an elephant, but it didn't matter. His brother had, once. That was enough.

The universe for people living closely with the land comprises of what surrounds them, what affects their food and fields, be it African veldt or Brazilian forest. They live and die with the effects of biodiverse losses, as trees are logged and waters turn polluted. They *feel* as things change around them, but cannot comprehend why it must happen. An elephant is somewhere else, not here; today's dust storm or lost forest is real. There is no insulation from harm when you live with the land. I learned something that night, on damp earth to the west of the Great Rift.

If all life is a library, with each volume a species, too many shelves are already dusty and bare. Only the survivors of 630 million years of planet earth's evolution are now racked in the library, and corridors echo as we walk away. Each day books are withdrawn and thrown to the ground, and with them goes that knowledge. When will we, the members of industrial nations who insulate ourselves from our biodiverse legacy, awake to what we have lost? And continue to lose. We are once removed from the effects we create, we draw up plans to help other continents save energy, but burn our street lights as beacons into space. Then, squandering our resources, we draw more lifeblood from distant forests and oceans where we do not have to deal face to face with the consequences. Industrial man is a true hypocrite.

Leave aside the arguments for population control, genetic preservation and fears for a sickening Gaia. If there is no value to life, if life has no intrinsic right to exist, what is the use of even caring about extinctions? The insects, snakes and microbes descending to oblivion

The inquisitive Goeldi's Marmoset, *Callimico goeldii*, is a rare primate resident in the upper Amazon region of South America.

alongside the furry, warm mammals matter not one piece. You, who have read this book, perhaps feel otherwise. You have to convince others. Those who care must look to the interests of other species than our own, for in so doing we, the caretakers, may care for ourselves.

We are part of Gaia, whether we appreciate it or not, and those with power must exert it to benefit our biodiverse world. The power lies with all of us, not just faceless governments or commercial combines. A lone voice today, tomorrow becomes an essential chorus of need. In a Kenyan dawn, I can appreciate the wonders of life, and wonder with sadness at a past world now gone. Yet, today is only tomorrow's nostalgic past, and the ashes of biodiversity increase with every hour. Gaia's soul is in retreat; *we* have brought the goddess to this point. Now, time is indeed short. The future is with us.

Ndiyo. It is so.

Early morning mist surrounds the coastal redwood forest which dominates northern California, individual trees towering over 100 m from the forest floor – they dwarf the Statue of Liberty – as the tallest trees in the world. Giant redwoods are also amongst the most ancient of living things, in excess of 3,500 years old. Redwoods are resistant to all known fatal diseases, but 85 per cent of the original forests have fallen to man's greed for lumber.

GLOSSARY

acid rain: Rainwater which dissolves pollutant gases as it falls, becoming acidic. Typically, the pollution is sulphur dioxide, derived from burning fossil fuels such as coal, and producing sulphuric acid. The direct effects of acid rain (as opposed to other pollution) are difficult to measure, as pollution can also fall directly to earth from the air, and may react with minerals in the soil to produce other, more poisonous compounds.

adaptation: An inheritable change in a species, be it physical or behavioural, which improves its ability to survive and compete with other species within its environment. The change, randomly introduced under the theories of evolution, will only prove beneficial if it confers such an advantage through the trials of natural selection.

aerobic: A process in which energy is released from large molecules within a cell, and which requires oxygen. Most of earth's species are aerobic organisms.

allele: An alternative form of a gene, producing variation within a characteristic. For example, a gene codes for a flower colour, the allele representing red, white, yellow or some other colour.

anaerobic: A process in which energy is released from molecules within a cell, and which does not require oxygen. An example is the action of yeast during fermentation.

artificial selection: The deliberate selection of characteristics as a criterion for breeding, for example to obtain a specific colour in the next generation of flowering plants. The result is a breed of animal, such as a dog (poodle, terrier) or plant (tall or short peas). As the effects of natural selection are avoided, it is possible (and common) for undesirable characteristics to also be maintained; it is estimated that virtually all breeds of dog possess at least one genetic disorder.

base: *See* nucleotide base.

binomial nomenclature: A system of classification, introduced by Linnaeus, used to provide all species with a unique name. Based on Latin, the binomial system uses two names: the genus and species.

biodiversity: The biological variety of all organisms of all classifications, from all kingdoms, that are found on earth.

biomass: The biological mass of living material within an area, or of a species or individual specimen (whichever is required), usually measured as the dry weight and used as an indication of productivity (growth).

biome: A major ecosystem type, such as rainforest, savannah or tundra, classified according to its dominant, representative plant.

biosphere: The inhabitable portion of the earth: its surface, subsoil and atmosphere to the limits at which life can survive.

breed: A stock or type of organism existing within a species, e.g. types of dogs. All belong to the same species (can interbreed) and superficially appear similar, but each possesses obviously different characteristics. Breeds are usually the product of artificial selection as opposed to natural selection.

cash crop: A crop grown for sale rather than food. It is sometimes more economic to use land for a cash crop, because a greater quantity of food can be purchased with the results of the sale than could be grown in the same area under the same conditions. Sources of drugs may also be considered cash crops. Examples of cash crops therefore include coffee, tea and opium.

cell: The basic building unit of living things. An animal cell is a mass of material surrounded by a membrane; a plant cell has an additional, exterior cell wall made of cellulose.

Cenozoic era: An era of time that began 65 million years ago and is subdivided into the Tertiary and Quaternary periods.

characteristic: A physical or behavioural attribute in an organism or species. This may be acquired during life (such as body weight being high or low due to diet or disease) or inherited (such as the colour of fur or eyes). Variation in characteristics in a species forms the basis of the evolutionary force of natural selection.

chemosynthesis: The release of energy for respiration from a simple chemical compound, such as the oxidation of hydrogen sulphide to sulphur. The process is typically used by some species of bacteria.

chromosome: A DNA and protein strand which occurs in pairs (one donated from each parent) in the nucleus of each cell, or singly in reproductive cells such as sperm, ova and pollen. In cells which do not bear a nucleus (blue-green algae and bacteria) the thread-like strand exists as a circular chromosome. The chromosome incorporates the organism's genes.

class: A group of similar orders used in the process of classification, a group of similar classes being collectively termed a phylum.

classification: A system of placing organisms into groups, each group possessing common characteristics. The first group, termed the kingdom, comprises several phyla and descends through the class, order, family, genus and species. The species therefore represents a unique type of organism. As an example, the animal kingdom contains the phylum Chordata, which contains the class Mammalia, the order Carnivora, the family Felidae, the genus *Felis* and the species *sylvestris*. The Latin name of *Felis sylvestris* represents the Wildcat. *See also* binomial nomenclature.

clone: A member of a species produced from a single parent, and therefore genetically identical to the parent.

cold-blooded: A cold-blooded animal (i.e. all groups except mammals and birds) is incapable of regulating its body temperature; it is directly influenced by its surrounding environment. Temperatures can be controlled by behavioural means, such as basking in the sun or moving into the shade. The term cold-blooded is therefore a misnomer; it is perfectly possible for a cold-blooded animal (properly termed poikilothermic) to be hotter than a warm-blooded one under certain conditions. *See also* enzyme.

community: Individuals of different species living within a habitat, which in turn forms part of an ecosystem. Interactions occur through such factors as food webs and in competition for space.

competition: Where resources are limited, there is competition between individuals of the same or different species. Competition is an essential part of the theory of evolution, being a force which leads towards the formation of new species through natural selection.

continental drift: The slow movement of continental-sized portions of the earth's crust across the surface of the earth. This causes slow changes in climate and terrain for resident organisms as the land masses pass through the tropics or over the poles.

cross: When two parent plants reproduce, they are said to be crossed; cross-pollination refers to the transfer of pollen from the anther of one flower to the stigma of a flower belonging to another plant of the same species.

deciduous: A plant (usually a tree) which loses its leaves at one time of the year, usually in the autumn. These are replaced with new growth in the spring. *See also* evergreen.

desert: An area of land which receives less than 25 cm of rain a year. Typically thought of as sandy, infertile wastes, this is not necessarily the case for all desert terrains.

DNA: Deoxyribonucleic acid, a long-chain chemical consisting of nucleotide bases connected in a chain. DNA is the material, twisted into the shape of a double helix, which is passed from parents to offspring and controls inheritable characteristics. DNA specifies the sequence of the 20 types of amino acids used to synthesize protein, each unit of the coded language consisting of a group of three nucleotide bases which form a gene. Strands of DNA are not constrained in bacterial and blue-green algal cells, but are enclosed inside the nucleus of all other cells. *See also* gene *and* replication.

ecology: The study of animals and plants and their interrelationships within their community and environment.

ecosystem: A community of animals and plants and their environment, e.g. a rainforest or desert ecosystem.

environment: The factors which comprise an organism's surroundings, such as terrain, climate, other species and food supply.

enzyme: An enzyme is a large protein molecule with a specific shape. This shape enables it to lock onto other molecules, facilitating chemical reactions which cause the molecules to join together (as might occur during growth) or split apart (digestion or respiration). Once the reaction is completed the enzyme remains to act upon other chemical substrates; enzymes therefore act as catalysts, speeding up chemical reactions. To operate efficiently, enzymes require suitable alkaline or acid conditions, the presence of water, and a suitable temperature. Their operation is therefore directly affected by whether the organism is warm- or cold-blooded; the former's chemical reactions may occur at any time, while the latter's depend on suitable environmental conditions to raise body temperatures to the correct level. Too much heat destroys the effectiveness of an enzyme, as the molecule's shape is changed and it can no longer perform its function. Boiling an egg has the same effect; the egg becomes solid as its proteins change shape.

ethnobotany: From 'ethnic' and 'botany': the study of plants and their (undocumented) uses by ethnic groups.

eukaryote: One of two types of cells, structurally containing a nucleus enclosing genetic material. This eukaryotic cell is found in all organisms with the exception of bacteria and blue-green algae. *See also* prokaryote.

evergreen: A plant which bears leaves all the year round; the common definition of 'doesn't lose its leaves in winter' is false as leaves are lost and replaced throughout all seasons. *See also* deciduous.

evolution: As propounded by Darwin, evolution is the result of a process which leads, via successive minute changes based on the results of competition, to the formation of new species. In the struggle for survival while competing for resources, some variations within a species will prove beneficial. These individuals will therefore possess an advantage, and be more likely to pass their characteristics on to the next generation. This sequence is one of natural selection. Additionally, there may be sudden changes in the genetic structure of an organism (a mutation) which introduce new variations for the evolutionary process to 'test' for any inherent, useful advantage or disadvantage.

exoskeleton: An external skeleton used for support and protection, for example that of insects and crabs. The alternatives are the possession of an endoskeleton (internal support, such as the bones of birds and mammals) and a hydroskeleton based on fluid (worms). Exoskeletons present difficulties for growth, and the external structure forces this to take place in spurts, each being punctuated by the old 'skin' or carapace being shed.

extinction: The total loss of all members of a species, due to a death rate which is higher than the birth rate. When this occurs over a short period of time it is termed a mass extinction, estimated to account for 60 per cent of species loss in the past. The steady rate of loss, spread through the geological past, forms the background rate of extinction. Species may also evolve into new forms, thus effectively making their ancestors extinct: a 'pseudo' extinction as, though the species is gone, its genes continue.

family: A group of similar genera, used as part of the classification of organisms. A group of similar families collectively form an order.

fertilization: The fusion of the nuclei of two reproductive cells: pollen and an ovum in seed-bearing plants, or spermatozoa and an ovum in animals. The process enables a new mixture of genetic material to be produced.

food chain: A chain of organisms which passes energy derived from food from one individual to the next. The number of links in the chain is not normally more than five, and usually fewer, as only some 10 per cent of the energy is utilized at each stage. Food chains begin with green plants (unless the chain is a chemosynthetic one), and proceed via herbivores to carnivores; they can also be built up into food webs. These show the relationships between all organisms involved in the chains, indicating every part of the species' diet.

fossil: The preserved remains of an organism or its imprint, such as a footprint, in sedimentary rock.

Gaia: A theory concerning the earth's self-regulating ability, where imbalances are corrected by feedback mechanisms. These are natural systems, with no implication of guided action.

gamete: A reproductive cell: an ovum (plural: ova), sperm or pollen. Gametes contain half the 'normal' number of chromosomes, consisting of a single rather than a paired set. During reproduction two gametes combine to form a single cell, the zygote, which therefore possesses a full complement of paired chromosomes (one set from each gamete, supplied by each parent).

gene: DNA comprises a series of coded instructions which enables it to control the correct formation of protein molecules. Each piece of code is a codon, which is built of a sequence of three nucleotide bases. A sequence of codons comprises a gene. The gene is inheritable. It is the intermix of genes passed from two parents to their offspring, and the presence of possible mutations, which determines the characteristics of the next generation and therefore introduces variation into the species.

gene map: A representation of a gene's position on a chromosome.

gene pool: All genes present in a population. *See also* genome.

genera: *See* genus.

genetic drift: Any population of individuals of a species will exhibit variation in its characteristics. A change in the total genetic composition of the population, due to random, chance factors, is termed genetic drift. In small populations this could cause the total loss of genes for some variations and a permanent alteration to the gene pool.

genetic engineering: A deliberate alteration to the genetic structure of DNA by the process of transgenics (transfer of genes from a source to a target cell), directly altering the DNA or introducing new DNA from another source. For example, this could involve the addition of a human gene into a bacterium so that the host bacterial cell produces insulin.

genome: The total of all genes present in an organism or species. The Human Genome Project is concerned with mapping the position of all human genes as well as their base sequences (the sequence of nucleotide bases which form the genes) in one person. The Human Genome Diversity Project is concerned with determining the differences between the genetic composition of all human ethnic groups.

genus: A genus (plural: genera) is a group of similar species which have characteristics in common and which are distinct from other groups. Similar genera are grouped into a family.

Gondwanaland: A major land mass which, due to continental drift, helped to form the supercontinent of Pangaea. When Pangaea broke apart, Gondwanaland continued its drift and later broke into smaller fractions to help form the continents of today's southern hemisphere.

greenhouse effect: A warming effect in the earth's atmosphere, due to the presence of increased gases such as carbon dioxide. Greenhouse gases are transparent to the sun's radiation, which therefore reaches the earth, but the gases trap heat and prevent it escaping back into space. The increase in heat is often referred to as global warming.

habitat: A portion of the environment which is occupied by an organism. The concept includes other individuals of the same and other species, as well as physical conditions such as soil type and climate.

hot spot: A region which is extremely rich in species and is also under threat, for example the Amazonian rainforest and Madagascar.

hybrid: The offspring of two parents belonging to different species (or, at a minimum, two parents which are genetically different, such as might occur when speciation has begun but not yet proceeded far enough to allow the two parents to be recognized as separate species or sub-species). The offspring are normally, though not always, sterile. Examples of species able to interbreed include a donkey/ass and a horse to produce a mule/hinny, and the Cowslip and Primrose to produce an Oxlip. *See also* polyploid.

immune system: Organisms contain a naturally occurring immune system used to defend themselves against invading organisms. Any foreign matter within the organism triggers the immune system to combat it, a problem with organ transplants as it leads to rejection of the donated organ unless its tissue is an extremely close match for the host's.

island: Ecological islands are similar in nature to those surrounded by water; they are isolated habitats which are cut off from other, similar habitats by the surrounding ecosystem. A pond forms an ecological island if there is no viable means of its organisms being able to interchange with other islands of the same type (in this instance, another pond). The isolation therefore enables evolutionary pressures to produce new species which are more ideally suited to the specific, local conditions.

isotope: An element which exists in different forms based on their atomic mass, each of which is a different isotope. The chemical properties of the isotopes are the same. Two stable isotopes of carbon, carbon-12 and carbon-13, occur naturally in the biosphere, the lighter (carbon-12) accumulating in organic material in preference to the heavier (the numbers 12 and 13 respectively, refer to the number of nucleons in the nucleus: the protons and neutrons). The half-life of the radioactive carbon-14 is used to measure the age of relatively recent organic matter; as it decays towards its lighter forms, the amount of carbon-14 present in the material decreases at a known, fixed rate.

K/T boundary: The junction between the rocks of the Cretaceous and Tertiary periods, referred to using the letters which represent these geological periods: K for Cretaceous (from the German *Kreide*, meaning chalk) and T (Tertiary). At the boundary a higher than expected proportion of iridium metal is found, indicating the possibility of a 10 km diameter comet, asteroid or meteor having collided with the earth and triggering the mass extinction of the dinosaurs.

kingdom: The highest rank used in the classification of organisms, consisting of five groups: Plantae, Animalia, Monera (bacteria and blue-green algae), Protista (some single-celled algae and the former category of single-celled animals, the Protozoa) and Fungi. Within each kingdom, groups with increasing similarities are found, classified into phylum, class, order, family and, finally, genus and species. *See also* classification.

limiting factor: When organisms grow they require raw materials; photosynthesis uses carbon dioxide, water and sunlight energy to produce sugars, for example. Any raw material which comes into short supply (such as carbon dioxide in a rainforest, water in a desert, or light with increasing depths in the ocean) will cause photosynthesis to stop, even if the other materials are present in abundance; it is a limiting factor.

Mesozoic era: An era of time which began 250 million years ago and is subdivided into the Triassic, Jurassic and Cretaceous periods, the latter ending 65 million years ago.

micro-organism: An organism so small that a microscope is required to view it. This includes single-celled plants and animals, but typically is applied to bacteria. Also referred to as a microbe, and incorrectly as a germ; germs are those micro-organisms which man defines as causing disease. Microbiology involves the study of micro-organisms.

molecule: A group of atoms bonded together into a single unit. For example, carbon dioxide (CO_2) consists of one carbon atom and two oxygen atoms and forms a single molecule. Some molecules are extremely large, e.g. protein molecules.

Monera: One of the five kingdoms of classification, consisting of bacteria and blue-green algae. Cells in this kingdom do not have a nucleus to enclose their strands of DNA.

mutation: A change in the genetic structure of an organism, resulting in the production of new characteristics in its descendants. Mutations are an essential part of evolution, some new characteristics being beneficial while others are harmful: only the former will survive the tests of natural selection. The mutation, which is inheritable, can be induced by such agencies as X-rays and some pollutants.

natural selection: The mechanism of selection of characteristics in an organism, leading to evolutionary change. Variations arise by random chance, so that the selection process is based on characteristics which may be useful, harmful or neutral to the organism's existence. The selection process is therefore positive for 'useful' characteristics (e.g. better camouflage in a prey species) and more likely to be passed on to the next generation as the organism has a better chance of survival and reproducing. *See also* artificial selection.

neo-Darwinism: A modified form of Darwin's theory of evolution, taking on board the knowledge gained from modern research into related fields such as genetics and ecology.

niche: The specific surroundings of a species, including aspects of its food and habitat. Thus, a pond habitat may contain many niches where species survive, such as under a rock or on a plant leaf. As species evolve they adapt to a niche; new adaptations may therefore permit a new niche to be utilized.

nucleotide base: There are four possible nucleotide bases (often simply referred to as 'bases'). Each is a chemical which, used in groups of three, forms the genetic code in strands of DNA; genes are therefore composed of nucleotide bases. Using three bases means a maximum of 64 different combinations can be formed, more than enough to control the 20 amino acids and allow for 'stop' and 'start' instructions.

nucleus: The portion of a cell which contains DNA in the form of chromosomes.

order: A group of similar families used in the process of classification, a group of similar orders being collectively termed a class.

organic chemistry: The study of carbon compounds (as opposed to inorganic chemistry, which is the study of non-carbon compounds that are usually based on minerals). Carbon is the essential element upon which life is based. Organic material therefore includes any material which is, or once was, alive.

organism: An individual member of a species.

Palaeozoic era: An era of time, subdivided into six periods. The oldest is the Cambrian period, which began 570 million years ago. In order of age, the next oldest is the Ordovician period, followed by the Silurian, Devonian, Carboniferous and finally the Permian period, which closed 250 million years ago. Major extinctions occurred at the end of the Palaeozoic, including the trilobites and other invertebrates.

pandemic: A disease which is prevalent across a large area, perhaps a country or the whole world. An epidemic is a widespread disease within a community, generally of a more restricted nature than a pandemic.

Pangaea: A super-continent which formed in the Permian period over 250 million years ago, to break up once again in the Cretaceous period, which ended 65 million years ago. Both occasions coincided with major extinctions, the former when the trilobites died out, while the latter was at the end of the age of dinosaurs.

parasite: A species which lives on or in another, living species (the host) to the host's detriment.

parthenogenesis: Reproduction based on the development of an unfertilized egg, producing genetically identical clones of the parent.

photosynthesis: The production of sugar by green plants using the raw materials of carbon dioxide, water and sunlight energy. Plants are therefore referred to as producers; food chains rely totally upon this production, which releases oxygen as a waste product back into the air.

phylum: The phylum (plural: phyla) is a group of similar orders, used as part of a system of classification. The phyla are grouped together to form a kingdom.

plankton: Plankton consists of minute plants (phytoplankton) and animals (zooplankton) which drift or swim in the sunlit upper reaches of the ocean or other bodies of water.

plasmid: A circular piece of DNA found within bacterial cells. Plasmids can be removed and spliced with genes from other cells before being reintroduced to the bacteria, making them important in genetic engineering.

pollination: The transfer of pollen from a flower's anther (the male organ) to the stigma (the female organ). This may be accomplished by wind or animals. Cross-pollination requires flowers belonging to two individual plants and, as this enables an intermix of genes, is the preferred method. Self-pollination may occur when cross-pollination has failed, producing identical seeds to the parent plant. *See also* fertilization.

polyploid: A cell containing an unbalanced selection of chromosomes; i.e. not all the chromosomes can exist as pairs. This may arise when two similar organisms from different species interbreed, or when the process of cell division (when chromosomes are replicated) is 'faulty', producing an extra set of chromosomes. A polyploid species is normally sterile, as reproductive cells cannot form with half the 'normal' number of chromosomes. The situation is common in plants. In animals, however, this is not possible and polyploid species are rare. *See also* hybrid.

population: Organisms of the same species living together, usually deemed to be separated from other groups of the same species by some means, enabling the forces of evolution to have an effect.

prokaryote: One of two types of cell, structurally containing genetic material in the form of filaments of DNA without a nucleus. This prokaryotic cell is found in bacteria and blue-green algae. *See also* eukaryote.

protein: A large, complex organic compound composed of amino acid molecules. Used for growth, for example to form muscle in animals, protein also forms enzymes.

Protista: One of the five kingdoms of classification, consisting of single-celled organisms (including those known as the Protozoa).

rainforest: A tropical forest with over 100 cm of rainfall a year, though areas such as the rainforests of

South America may receive in excess of 200 cm of rainfall a year. Unlike a desert (with 25 cm of rainfall a year, which may all fall during a limited period of time), rainforests receive rainfall throughout the year. Typically, growth of trees is rapid and the area is extremely biodiverse.

replication: The production of an exact copy of DNA. This is shaped as a 'double helix': a twisted and then recoiled double-stranded molecule. During replication the strands separate and reconstruct a new strand for each half, identical to the old. The end result of this manufacturing process is two, identical double helixes of DNA in place of the original one.

respiration: The release of energy from organic material ('food') by the cells of living organisms, producing water and carbon dioxide as waste products.

retrovirus: A virus which contains a single strand of genetic material based on RNA. It is used in some genetic engineering operations to transfer genetic code, as retroviruses will incorporate their genes into the host's DNA, which is replicated each time the host cell reproduces.

RNA: Ribonucleic acid. RNA exists in three forms, acting to transfer information from DNA to the site of protein manufacture in the cell. RNA is therefore essential to ensure the codes contained in the cell's DNA are correctly carried out, producing specific proteins for growth. Retroviruses use RNA only rather than the DNA/RNA combination.

speciation: The process of the production of a new species, due to evolution.

species: A unique form of life which is able to interbreed only with other members of the same species; the final group used in classification with similar species comprising the genus.

symbiosis: Two species living intimately together, each benefiting from the close association. Examples include the zooxanthellae algae found within corals, and algae living within fungal structures to form lichens.

taxonomy: The process of classification of organisms, both living and extinct.

transgenic techniques: 'Transferred gene' techniques, e.g. where a gene from one species is introduced into the DNA of another, causing a change in the genetic structure. *See also* genetic engineering.

transpiration: Land plants lose water by evaporation during the process of transpiration. As a result, more water is drawn through the plant, forming the transpiration stream.

virus: A microscopic, parasitic form of life that consists of little more than a strand of DNA which the virus inserts into a host cell, causing it to produce more viruses.

volcanic hot spot: An area of high volcanic activity, existing now or in the past. One theory indicates that meteorite strikes could have initiated the formation of volcanoes and caused lava to flow into the ensuing meteor's crater (forming basalt lakes), covering the evidence. Dating basalt flows has produced good correlation with the dates of mass extinctions. There is confirming evidence that the last mass extinction, 65 million years ago, was indeed triggered by a collision with a large meteorite or comet.

warm-blooded: Mammals and birds are warm-blooded, a condition properly referred to as homoeothermy. As such, body temperature is maintained at a constant level, usually above the organism's surroundings. *See also* enzyme.

zooxanthellae: Algae found within coral polyps in a symbiotic relationship. The coral benefits from the products of photosynthesis supplied by algae, and is forced to maintain a position in bright, sunlit, warm waters for the algae's benefit. There are therefore similarities in structure between corals and plants, in their attempts to present a large surface area to sunlight for maximum absorption.

INDEX